62047

Design Guidelines
for
Surface Mount
Technology
Vern Solberg

TAB Professional and Reference Books

Division of TAB BOOKS Inc.

Blue Ridge Summit, PA

Published by **TAB BOOKS Inc.**
FIRST EDITION/FIRST PRINTING

Library of Congress Cataloging-in-Publication Data

Solberg, Vern.
   Design guidelines for surface mount technology / by Vern Solberg.
     p. cm.
   ISBN 0-8306-3199-2
   1. Printed circuits—Design and construction. 2. Surface mount
technology. I. Title.
TK7868.P7S638 1990
621.381'531—dc20
                                         89-374799
                                           CIP

TAB BOOKS Inc. offers software for sale. For information and a catalog, please contact TAB Software Department, Blue Ridge Summit, PA 17294-0850.

Questions regarding the content of this book should be addressed to:
Reader Inquiry Branch
TAB BOOKS Inc.
Blue Ridge Summit, PA 17294-0850

Front cover photograph courtesy of SCI Manufacturing, Inc. Design Center, San Jose, California. The photograph shows a board created for Hycom Inc., Irvine, California.

Vice President and Director of Acquisitions: Larry Hager
Production: Katherine Brown
Series Design: Jaclyn B. Saunders

Contents

Dedication:
To my very supportive and patient wife, Roi.

Acknowledgments

My gratitude and thanks to the many associates who have supported my efforts in testing and documenting the material in this book. A special thanks to Mike Brisky, Steve Dow, Norbert Socolowski, and Phil Zarrow.

Preface

Surface-mounted, or surface mount, technology came into its own in the 1980s with the rapid evolution of robotic assembly equipment, improved solder technology, and a greater supply of miniature SMT components.

This was not always the case. As early as 1983, many people were still saying "Surface . . . what?" and I found myself hand-sketching diagrams of how SMT worked because at the time, there was no documentation and few standards, and how-to books on the subject were nonexistent.

The need for reference manuals or guidelines for designers using surface mount devices was apparent. I gathered up the sketches and rough illustrations—many on the back of envelopes or odd scraps of paper—and in 1983 the original *Design Guidelines for Surface Mount Technology* was put together. It was the first of its kind in the electronics industry.

Then, as now, we tested the component land patterns and design rules through actual in-process use before including them in the book. Since 1983, the technology has grown rapidly, and this book includes the new developments in components and materials.

The rapid growth of SMT has been the result of innovation, continual process development, testing, and implementation of new ideas. I am pleased to have been a part of the early development of the technology and have enjoyed work with various groups, including IPC Task Force on Design Standards, SMTA, SMT EXPO, NEPCON, EIA,

IEEE, and currently with the IPC Advanced Packaging Committee to update the *Electronic Packaging Design Handbook*. Companies throughout the United States have adapted the guidelines presented in this book as the basis for their in-house SMT standards.

Through the years of working in SMT, the one fact I've found that holds true is this: the success of SMT product manufacturing depends on a good design!

Introduction

There are no secrets to successfully implementing surface-mounted technology in your electronic products—only process proven techniques.

Design Guidelines for Surface Mount Technology was written to prepare the printed circuit designers with many of these techniques for developing circuit substrates. The book emphasizes why today's professional must be aware of the relationship between substrate design and the manufacturing process, since the design often influences the success of the process.

The text and illustrations will provide you with step-by-step procedures in developing the most cost-effective product possible using SMT, including alternative methods of packaging, substrate selection, and fabrication options.

Written by a designer for designers, the detailed information is useful to several disciplines, including printed circuit designers, manufacturing engineers, program managers, and quality engineers, as well as anyone involved in developing or manufacturing electronic circuit assemblies.

Designers having experience using more conventional pin-through-hole (PTH) leaded components will also find the book beneficial. It is a common practice to mix leaded and surface-mounted devices on the same substrate. This mix of technology (SMT/PTH) or *hybrid*, can be adapted to any substrate type, both rigid and flexible.

While computer-aided design (CAD) has proven very efficient in SMT design, it is not mandatory in order to implement and lay out SMT circuits. Land pattern standards and methods for developing the contact geometry for nonstandard devices are clearly illustrated. Recommendations for component selection, placement guidelines, and minimum spacing of components are detailed with tables to assist in calculating component density.

Through illustrations, design guidelines are furnished for all types of SMT applications, including examples of preferred layout and methods used for signal routing of high-density multilayer circuits.

Design Guidelines for Surface Mount Technology is a workbook, not an overview. Each chapter has been written, illustrated, and detailed to guide the user in the implementation of advanced, process-proven techniques for SMT assemblies, rather than untried or unproven theories.

1

SMT Layout Guidelines for Rigid Circuits

URFACE MOUNT TECHNOLOGY (SMT) IS A MANUFAC-
turing process that, through miniaturization, provides increased
component density for electronic products. This process offers
the user a means of producing, through robotic assembly, high-volume
products with a minimum of labor.

In this chapter, design guidelines are furnished for the layout and
interconnection of the SMT circuit board. These recommendations have
been process-proven. The designer adhering to these guidelines will
benefit from the success of others.

Components, assembly equipment, and materials for attaching the
surface-mounted parts have improved as a result of this early work.
These segments of the industry have striven to standardize in areas that
will complement each other. Industry-sponsored organizations and gov-
ernment agencies have worked together in reviewing and approving
these standards.

When a component manufacturer introduces a new package to the
industry, it is usually a product that will benefit the user and it is a wel-
come addition to the complement of devices already in use. The component industry also will develop packaging that can be used or adapted to
a wide variety of assembly systems.

Newly developed assembly equipment having different functions in
the overall SMT process is made compatible through the cooperative
effort of the Surface Mount Equipment Manufacturers' Association
(SMEMA). In many cases, the working zones of different brands of
machinery are at or near the same height from the floor and will have
compatible conveyer systems with material flow and width adjustment
from a common direction. This compatibility allows the systems engineer to choose the best combination of equipment for a product's
assembly without compromising efficiency or precious resources.

COMPONENT SPACING FOR SMT

In Surface Mount Technology, component spacing is the key to successful assembly. Robotic assembly equipment places surface-mounted
devices (SMDs) on the substrate surface with an accuracy of $\pm.005$ to
$\pm.001$ inch of true position. To maximize component density, designers
are tempted to pack components of the PC board with only minimum
clearance between component bodies. The recommended guidelines
shown in this chapter provide space between components for ease of
inspection or rework.

The space between land patterns of the chip components shown in
FIG. 1-1 allows for a .010- to .012-inch-wide trace to pass between land

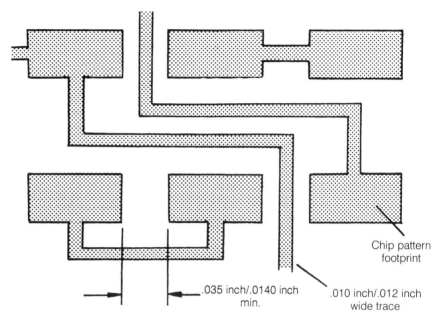

.035 inch/.0140 inch min.

Chip pattern footprint

.010 inch/.012 inch wide trace

Fig. 1-1. Land pattern spacing for surface-mounted chip components must allow for solder mask separation and circuit conductor routing.

patterns, while providing space for rework or touch-up tools. This spacing also reduces solder bridging when the components are to be wave soldered.

Low-profile IC component types such as the small outline and quad flat-pack ICs have gull-wing shaped leads protruding outward from the

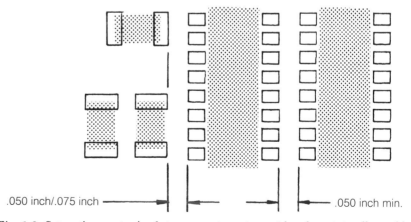

.050 inch/.075 inch

.050 inch min.

Fig. 1-2. Separation or spacing between components must be adequate to allow solder inspection and rework, when necessary.

body. This lead design accommodates inspection and accessibility for touch-up or rework. When space is available, separate these parts to provide clearance for rework or removal systems. When a surface-mounted IC is damaged, or for some reason must be replaced, there must be enough space between outlines to desolder and remove the faulty device without disturbing adjacent devices.

The component spacing as illustrated in FIG. 1-2 provides ideal accessibility when it is necessary to use rework or removal tools.

COMPONENT SPACING FOR
HIGHER-PROFILE COMPONENTS

The PC designer seeking maximum component density often mixes surface mount and leaded devices on the same substrate, but fails to provide adequate spacing between devices. Mixing high- and low-profile surface mount devices with leaded connectors and DIP ICs demands a greater attention to specific clearance guidelines. Spacing between the taller SMT components must permit visual inspection of the solder connections. A viewing angle to inspect the solder connection of a J-lead Plastic Chip Carrier (PLCC) requires a spacing of no less than .150 inch. This component spacing reserves access for solder touch-up or rework tools and test clip attachment. The same angle of view must be calculated when positioning taller molded capacitors or inductors near the high-profile ICs. It is common to mount DIP ICs parallel to one another with .100-inch spacing between lead rows. The solder connection to the lead- or pin-through-hole (PTH) allows for solder inspection on the side opposite the component body. When mixing the SMT and PTH devices in close proximity, the designer should plan for a viewing angle and rework tool access equal to the space recommended for the PLCC as previously noted. Mixed technology guidelines will be expanded later in this chapter.

Single In-Line Memory Modules (SIMM) assemblies are limited to a specific profile. The density of the SMT assembly occasionally dictates closer spacing than previously recommended. PLCC and SOJ devices are mounted with a minimum clearance as shown in FIG. 1-3. Some components are spaced as closely as .020 inch between lead surfaces. Visual inspection, in this case, is not possible and solder process verification can be made only with cross-section or x-ray evaluation. Both these methods require specialized equipment with skilled technicians, and generally inspection is on a random-sample basis.

When the SMT component density exceeds that previously recommended, the surface area opposite the traditional component side, or side one, can be used. Transferring the lower-profile discrete components to side two will reduce component density with very little effect on overall assembly cost. Chapter 5 will illustrate two-sided SMT assembly layout and clearance guidelines in greater detail.

Clearance between chip components must also accommodate inspection and rework tools. With adequate space, the danger of solder bridging and voids is eliminated. FIGURE 1-4 is an international guideline for component spacing for chip devices. The illustration includes the contact pattern and component body.

When a closely spaced surface-mounted IC fails and must be replaced on an assembly, the technician first cuts each lead from the body of the component before removing the part. Since access to the component footprint contact is restricted, and cannot be touched up, the entire assembly will be exposed to the solder reflow process again.

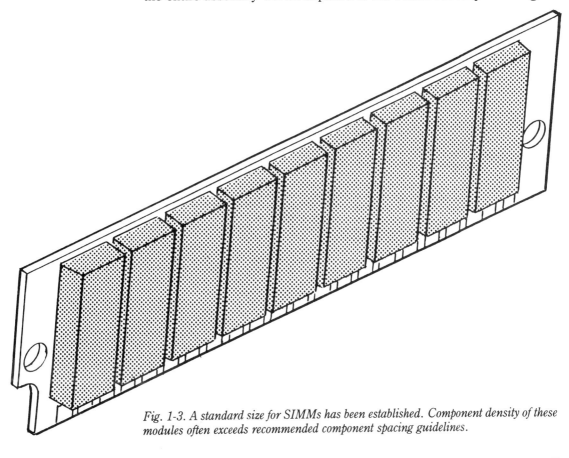

Fig. 1-3. A standard size for SIMMs has been established. Component density of these modules often exceeds recommended component spacing guidelines.

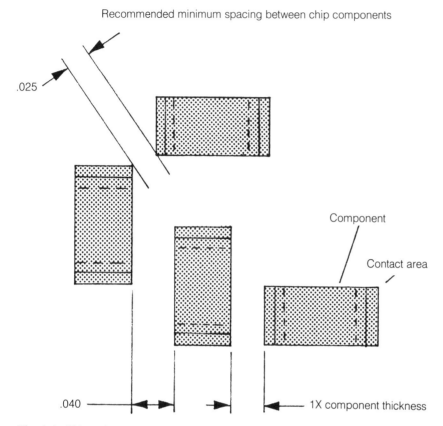

Recommended minimum spacing between chip components

Fig. 1-4. *Chip resistors and capacitors can be oriented in any direction, but the spacing between components must remain accessible to repair or replacement tools.*

TRACE-TO-CONTACT GUIDELINES

PC boards for SMT will have many variables that are dependent upon component density and interconnection complexity. Boards range from simple one- or two-circuit sides to the more complex multilayer laminated construction common in high-density PTH assemblies. The trace width as well as the air gap between traces and pads will effect quality and cost of the finished board.

Recommendations for circuit trace and air gap between traces for outer surfaces of a substrate should be .008 or .010 inch. When multilayer construction is required, fabricators will use either one ounce (.0014-inch-thick) or one-half ounce (.0007-inch-thick) copper-clad laminate. The inner layers, using one half ounce copper, can have trace width/air gap dimensions of .006 inch and less; but .008 inch will have a higher yield and will curb excessive fabrication cost.

Trace width of .003 inch with equal air gap is an everyday occurrence for some fabricators, but remember, high-tech boards are closely associated with high cost. The mainstream quality shop doing multilayer PC boards will recommend a more conservative approach whenever possible.

Trace-to-trace and trace-to-contact patterns shown in FIG. 1-5 allow for consistent etch quality on outer layers, while ensuring thorough coverage by the solder mask over the circuit trace. An exposed edge of the conductor passing near or between contact areas will attract solder particles and result in bridging. Removal of the component is the only way to eliminate a short when it occurs on an IC contact that folds under the component (J-lead PLCC).

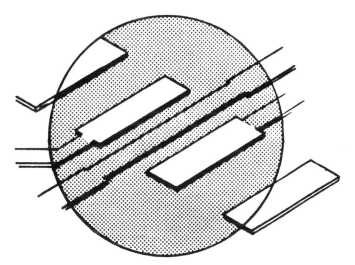

Fig. 1-5. Routing a .008-inch conductor between .050-inch spaced land patterns is a common design practice. The circuit trace must be covered with solder mask.

The use of narrow traces to interconnect the wider conductor to the component contact area is a subtle method of reducing the need for rework. If a very wide trace or copper area is connected to the contact pad by an equally wide trace, two reactions are possible during reflow solder. First, the component could be drawn off the liquid solder, since the solder contacts adjoining the large metal masses cool the solder quickly. In the case of chip components, one end of the component lifting away from the pad (*tombstoning*) is a common defect. The second reaction on tin-lead plated PC boards is a migration of the liquid solder paste during reflow processing. The liquefied solder paste will flow under the solder mask coating and merge with the plating on the conductor traces. This migration from one or more leads or contacts of the surface-mounted device will require additional solder touch-up. FIGURE 1-6 illustrates a few Dos and Don'ts to avoid solder problems.

Separation of the footprint patterns of two chip components will ensure containment of the solder paste. When it is important to increase conductor width between components, the best result is achieved with two narrow traces rather than one wide one. Containment of the solder within the footprint patterns is the key to controlling the solder process. (FIG. 1-7.)

Tin-lead solder plating of copper traces on circuit boards is common, but the alloy will return to a liquid state during the high-temperature reflow-solder process. While liquid, the solder will redistribute under the mask coating, resulting in an irregular appearance. When the molten material collects in a mass, cracks or breaks in the solder mask coating will occur.

The preferred circuit board for surface-mounted assembly will have solder mask over bare copper (SMOBC). The solder plating is limited to the exposed contact area of the components, while conductor traces and other copper plating under the mask material remain flat.

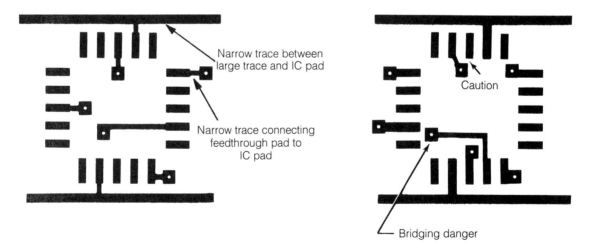

Fig. 1-6. Plated through-holes (or via holes) and pad area must be separated from the land pattern by solder mask in order to contain the solder paste during the reflow-soldering process.

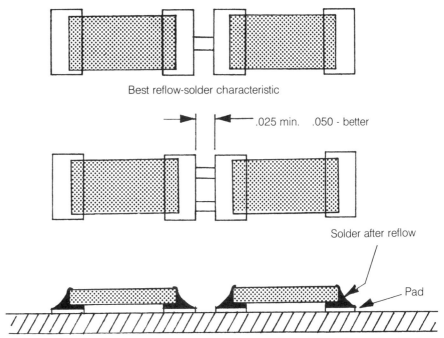

Best reflow-solder characteristic

.025 min.   .050 - better

Solder after reflow

Pad

Fig. 1-7. Spacing between components that are connected end-to-end must provide for solder mask separation of SMT land patterns.

CHIP COMPONENT FOOTPRINT OPTIONS

A footprint pattern that is too large for the chip component will encourage an excess of solder buildup. To reduce unwanted reworking, match the component to the proper size pad geometry. When wave soldering discrete components, you can adapt an optional narrow pad geometry to limit the amount of solder on the component connection. Be aware of the component body when positioning the footprint pads. If the narrow pattern is used, it will appear that the designer has adequate clearance, but when a component is placed, the potential for solder bridging to adjacent component leads increases. (FIG. 1-8.) Tolerances on component size, placement accuracy, and the PC board must be taken into account, as these tolerances can add up to ± .010 inch, or greater, from true center position.

SOLDER MASK FOR SOLDER CONTROL

Solder mask, a valuable ally in reflow assembly of surface-mounted products, can be applied with either a coat of a dry film photo-imaged

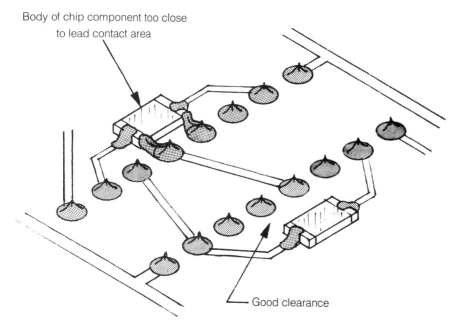

Body of chip component too close
to lead contact area

Good clearance

Fig. 1-8. Epoxy attachment of components for wave-solder processing is a common procedure. Spacing must be adequate to prevent solder bridging.

lamination process, or liquid photo-imaged polymer. Except for very small PC boards, the designer should always specify a photo-imaged mask coating.

Clearance of a photo-imaged solder mask can be zero to as great as .005-inch or .010-inch overall expansion. This clearance can be controlled by furnishing an expanded pad master to the board fabricator with instructions not to expand further.

The two examples shown in FIG. 1-9 compare the ideal solder mask clearance to the unacceptable. Solder mask options will be studied further in chapter 8.

When the wet screen pattern application of solder mask overlaps the contact area or when solder mask residue is left on the same area, reflow soldering will not be satisfactory. Likewise, when too great a clearance is allowed, the liquid solder will spread away from the contact area, promoting an unreliable solder connection.

CONTACT (FOOTPRINT) TO VIA PAD

The space between a *via pad* or feedthrough hole and the component contact area should provide for a solder mask barrier. This barrier

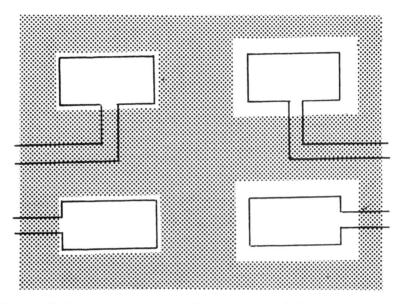

Fig. 1-9. Solder mask clearance around the contact must be kept to a minimum. A wide opening will allow solder paste migration when the alloy is in a liquid state.

will contain the liquid solder paste in the contact area during the reflow process.

The configuration shown in FIG. 1-10 provides a .008- to .010-inch-wide solder mask strip separating the contact from the via or feed-through pad and hole.

Fig. 1-10. If the space between the component land pattern and the via pad is too close for solder mask separation it may be necessary to cover or tent over the via hole and pad.

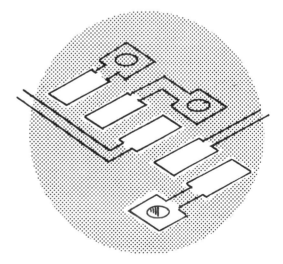

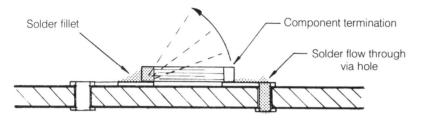

Fig. 1-11. Solder bridging under a chip component can be avoided by covering a via pad with solder mask or relocating the via to a clear area outside the device's body.

When a contact footprint must be joined to a via pad without adequate space, it will be necessary to cover the via pad with solder mask.

Note: If the via hole is too close to the component and is not covered with solder mask, the liquid solder will flow down through the hole and away from the intended component contact.

Feedthrough pads against or within the footprint pattern will also cause migration of liquid solder away from the contact area during reflow. Feedthrough pads under chip components cannot be seen, as shown in FIG. 1-11, and can cause failure during additional processes. Feedthrough pads and heavy traces adjoining the contact area will not have negative results in the wave-solder process.

A feedthrough pad under a chip device, as shown in FIG. 1-12, is not recommended. Solder or adhesives can migrate into the hole during the secondary assembly operation, thereby causing additional rework.

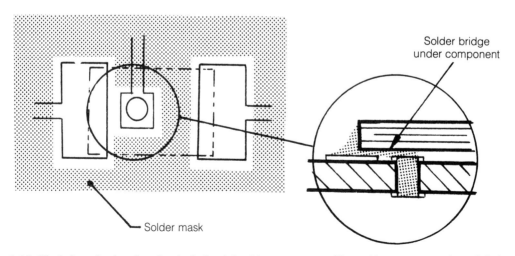

Fig. 1-12. Via hole pads placed under the body of the chip component will provide an unwanted conduit for solder migration during wave-solder processing.

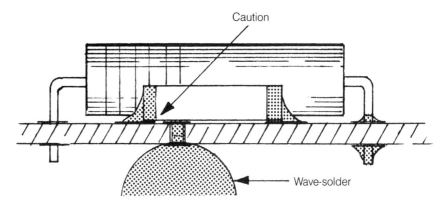

Fig. 1-13. Containment of solder paste on the component land pattern is vital to process control. When solder volume is not equal at each end of the chip device, separation or de-wetting may occur.

The migration of liquid solder will cause one end of the component to pull away from the contact area or flow into unwanted places during secondary wave-solder procedures. FIGURE 1-13 illustrates the migration of liquid solder through a closely connected feedthrough hole.

It is a common practice to cover or tent over via holes that are not test probe contacts with solder mask. The solder paste will be restricted only to the contact area of the device during the reflow-solder process.

ORIENTATION OF POLARIZED COMPONENTS

Equipment used to place polarized parts on the board's surface must be programmed to pick up, rotate, center and transfer each device to the desired location. With consistent direction and a symmetrical component arrangement, programming is simplified and inspection or rework is more efficient. The illustration in FIG. 1-14 is typical of what the designer can achieve when the layout is well planned.

Orientation and location requirements for automatic assembly methods are discussed in chapter 9.

MELF (TUBULAR-SHAPED) COMPONENTS

The recommended packaging for robotic assembly of MELF components, as with the chip components, is the tape-and-reel. When using the MELF diode, keep in mind the orientation of the component. FIGURE 1-15 illustrates the standard diode direction with the cathode end toward the index holes of the tape carrier. By maintaining consistent ori-

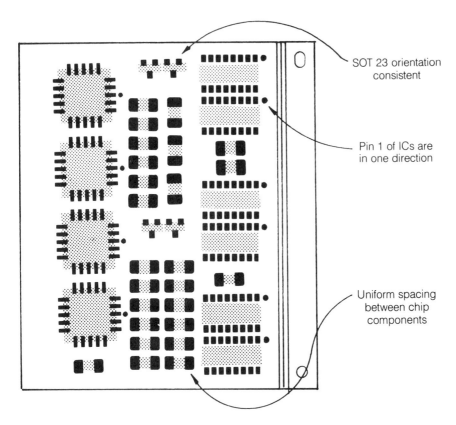

SOT 23 orientation
consistent

Pin 1 of ICs are
in one direction

Uniform spacing
between chip
components

Fig. 1-14. Maintain a consistent orientation of components when possible. The direction of components should be clearly defined on the substrate surface and assembly documentation.

entation of diodes, placement time per component can be minimal and inspection by QA more efficient. One direction is not always feasible but, when possible, it contributes to a proficient assembly.

INTERCONNECTING SOT-23 COMPONENTS

Reflow soldering of surface-mounted transistors (SOT) devices requires a minimum space between pad area contacts as compared in FIGS. 1-16 and 1-17. The distance specified in FIG. 1-16 furnishes an adequate solder mask area to prevent solder migration away from the component lead. Maintaining distance between contacts of the SOT device also reduces excess solder buildup, bridging, and voids when the wave-solder process is applied.

MELF diode orientation

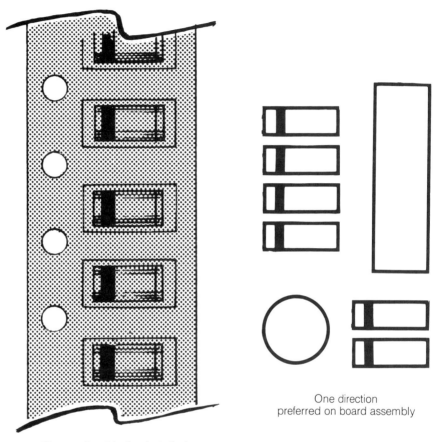

One direction
preferred on board assembly

Tape and reel (std. orientation)

Fig. 1-15. Common orientation of diodes and other polarized devices will reduce assembly errors and improve inspection time.

Fig. 1-16. The spacing provided for the SOT-23 land patterns and via pads will ensure solder containment.

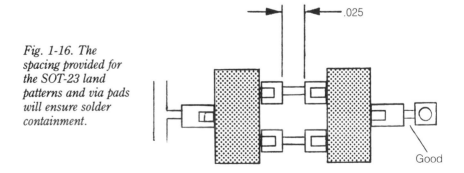

.025

Good

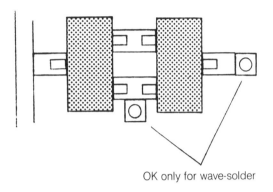

OK only for wave-solder

Fig. 1-17. Joining the contact land patterns of adjacent components and vias is not recommended for reflow-solder processing.

PLANNING FOR ASSEMBLY TEST

Every product requires testing before being passed on to the next assembly. A bench test is generally applied to low-volume or extremely sensitive assemblies. As volume increases, implementation of refined test methods will reduce labor-intensive hand-probing. Automatic test equipment is being developed and continually improved for SMT applications. Spring-loaded test probes are now available for .050-inch grid spacing. Because these probes are miniature, they are more delicate

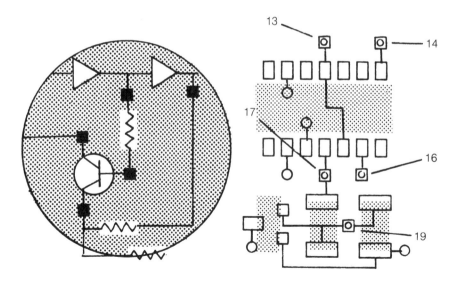

Fig. 1-18. The designer must provide for test probe access to every "net" or common connection between device contacts. Test probe contact is generally made on the surface opposite the side with the greater component population.

than the larger probes used on PC boards with leaded-through-hole technology.

As SMT assemblies become densely packaged, testing with miniature probes on dedicated fixtures becomes common. The quantity of test points required on each circuit will be determined by how refined the test system is and the type of testing required. On small modules in high-volume assembly operations, functional testing is popular. With more complex assemblies, it might require refined test methods. It is not advisable to rely on surface-mounted component lead contacts as test points. It is safer to use a feedthrough pad or add a contact area specifically for the point tested (FIG. 1-18). Test parameters are addressed in detail in chapter 6.

MIXED TECHNOLOGY—LEADED OR THROUGH-HOLE AND SURFACE-MOUNTED

Mixed technology—using leaded or pin-through-hole (PTH) with surface-mounted components—is often unavoidable. Until all components are available or economical for surface mounting, the need to continue using leaded-through-hole devices on the same substrate will remain.

CLEARANCE FOR LARGE PTH PARTS

Close attention to component spacing is important when the designer mixes PTH parts with SMT. PC board assemblies often require a header or a connector interface to another assembly or segment of the system. The connector products available in a surface-mounted configuration do not always have multiple sources. Identical SMT connectors from two or more manufacturers are not common, thus forcing the designer to revert to the leaded components. This is also true for the larger value capacitors, resistors or potentiometers found only in packages with leads.

The designer, looking at the layout in one dimension, may fail to consider the component height when positioning surface-mounted devices near leaded. The space between high-profile SMT and PTH parts must accommodate visual inspection and access of rework tools, soldering iron, test probes, etc. FIGURE 1-19 details the profile of the PLCC and the typical leaded through-hole device.

The visual angle on the J-lead part is just as important, as previously mentioned. Touch-up or rework tools require a minimum space to access the solder area without disturbing or damaging the adjacent component.

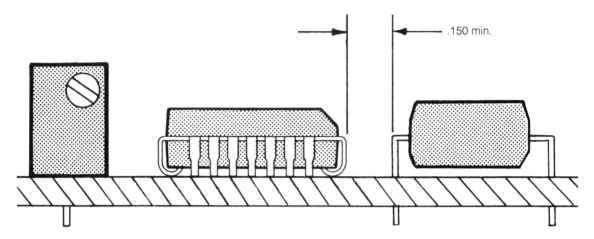

Fig. 1-19. When locating higher profile SMT components near other devices, reserve space for solder inspection and, if necessary, touchup tools.

COMPONENT-TO-BOARD EDGE REQUIREMENT

Surface-mounted devices can be mounted close to the substrate edge; however, the designer should avoid close edge placement, especially when the assembly is to pass through a wave-solder process.

When wave soldering is to be part of the assembly sequence, break-off strips on the long edges of the board will provide an ample holding surface for the machine's conveyer mechanism and protect components from possible damage.

FIGURE 1-20 is a typical application of a breakaway strip that includes tooling holes to be used with other machine handling. Generally, a board edge-to-component body clearance of .125 or .188 inch is preferred.

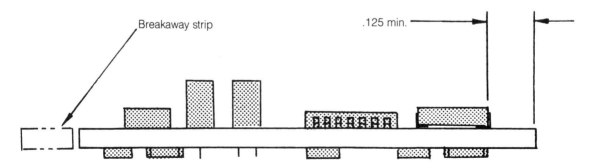

Fig. 1-20. Component body to substrate edge clearance must allow for direct machine handling. For those assemblies with a restricted edge clearance, a breakaway strip must be added to each unit.

In addition to the rigid substrate shown in this chapter, other substrate options are available to the designer. Custom-molded or structural substrates having conductor traces on one or more planes are possible. These three-dimensional products are prepared using conventional injection molding equipment, but with special plastic compositions. The materials were developed to remain stable and withstand solder reflow temperatures.

A more common adaptation of a well-established technology is the use of the flexible copper-clad polyimide materials (flexible circuits). Through incorporation of the components as part of the surface of the flexible substrate material, the assembly can conform to virtually any shape. Chapter 2 expands on the unique guidelines for developing the flexible SMT circuit.

2

SMT Layout Guidelines for Flexible Circuits

FLEXIBLE CIRCUIT INTERCONNECT SYSTEMS HAVE proven to be a rugged and reliable alternative to wire or cable interconnection methods. The materials used to fabricate these irregular shaped ribbons of copper onto non-rigid dielectric will conform to virtually any physical environment.

For SMT applications, the basic dielectric material must withstand temperatures required for various solder processes. The fabrication procedure for laminating and etching the flexible substrate or circuit is detailed in chapter 8. The base material primarily used for SMT is .002- to .003-inch-thick polyimide. The copper foil is one-half ounce to one ounce (.0007 to .0014 inch) thick. After the circuit pattern is etched to remove excess copper, an insulating layer or *coverlay* is laminated over the copper pattern. Only the SMT footprint pattern and other solder contact areas remain exposed.

In this chapter, guidelines are furnished to assist the designer in preparing the most reliable, economical, and process compatible product possible. Design guidelines for SMT on flexible circuits are, in many ways, the same as those shown in chapter 1; however, variations are necessary to accommodate the flexible circuit construction.

The design of the flexible circuit, although challenging, is far more versatile than a fixed-plane rigid circuit. The development and fabrication cost, when compared to rigid board circuit, will be greater. When comparing the unit cost, however, address the cost that is added to the rigid circuit through the attachment of wires, connectors, or mounting brackets. Most of these appendages can be incorporated into the one-piece design of the flexible unit—including selective rigid backing and mounting tabs.

The following text addresses five areas of flexible circuit design and fabrication options:

1. Conductor Trace Routing and Bend Radii Conventions
2. Trace Connections and Hole/Pad Filleting
3. SMT Footprint Guidelines and Coverlay Openings
4. Tear Restraints and Strain Relief Methods
5. Panelization and Fiducial Targets for SMT Automation

CONDUCTOR TRACE ROUTING

Single-sided flexible circuits have few limitations as far as conductor width and spacing. The notably critical factor is in the path of the circuit trace. A change in direction for a conductor will ideally have a gentle radius at the point of the turn. When using a CAD system for designing

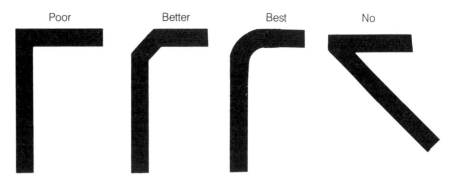

Fig. 2-1. Sharp corners of a conductor trace on a flexible substrate will cause tearing of the base material.

the flexible circuit, it is more common to provide a 45 degree turn as shown in FIG. 2-1.

Note: Do not use a 90 degree right angle turn. The sharp corner of the copper conductor promotes tearing of the base material.

Parallel trace runs should always be perpendicular to a ''fold line.'' Bending the copper on the bias is a poor practice that adds stress to the trace on the opposing side, which will damage the circuit (FIG. 2-2).

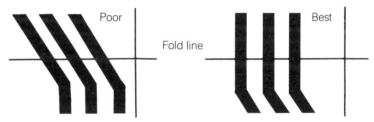

Fig. 2-2. If a bend in the base material is necessary, conductor paths should be perpendicular to the fold line. Folding the circuit traces across a bias direction should be avoided.

Two layer, parallel trace runs, depicted in FIG. 2-3, are offset or staggered from one side to the other. If conductors are directly opposite one another on a fold line, a stress is transferred to the trace as it attempts to stretch or compress around the bend.

TRACE WIDTH AND AIR GAP

Minimum trace width/air gap between conductor traces can be as little as .002/.002 inch on one-half ounce copper foil; however, a wider

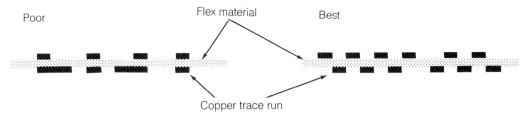

Fig. 2-3. If a fold line is required, conductor trace paths on two sides of the base material should be staggered from one side to the other.

trace and air gap dimension is desirable. A conductor/air gap width of .002/.003 inch, .003/.004 inch or wider, will give the fabricator a far better yield on the finished flex circuit, thereby maintaining a stable cost for the product.

TRACE CONNECTIONS AND FILLETING

Narrow conductor traces should taper or widen as they meet the larger SMT footprint or the mounting hole pad. The widening of the conductor adds physical strength at the interface junction to the larger shape. When joining trace and via hole on a flexible circuit, a minimum annular ring of .007 inch around the finished hole diameter is recommended; however, a .010 inch or greater ring is preferred. The margin will increase the bonding strength of the copper pad and trace to the base material to provide a more durable product. Fillet junctions can be tapered to blend a narrow circuit trace into a larger pattern or hole pad area. Not all CAD systems will accommodate this nonuniform geometry, and other transition lead-to-pad shapes must be created. A few examples are shown in FIG. 2-4.

One shape, popular with designers and fabricators, is the oval extension on the via pad. The oval pattern can be added to any contact area for greater strength when contained under the coverlay material or to provide for a trace entry connection. When space permits, simply extend or enlarge the contact area to allow for a perimeter or end entrapment of the copper material under the coverlay. This overlapping will add more strength and durability to the contact area and reduce the occurrence of separation of the copper foil when heated and re-heated by soldering tools or other processes.

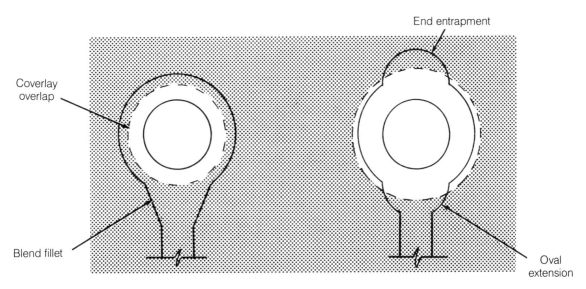

Fig. 2-4. The copper area around a nonplated hole will separate easily from the base material if not entrapped by the coverlay material.

SMT FOOTPRINT GUIDELINES AND COVERLAY OPENINGS

The flexible circuit footprint geometry for SMT will vary from the contact pattern used on conventional rigid board etched circuits. The PC board will generally have an epoxy or polymer coating applied as a solder mask with openings to expose the SMT footprint as shown in FIG. 2-5.

To insulate the conductor traces of a flexible circuit, the coverlay material is laminated to the flexible base circuit with openings furnished to expose the SMT contact patterns. These openings are prepared prior to lamination by die cutting or machining (FIG. 2-6).

The designer will modify the contact geometry as required to entrap the copper material under the coverlay for strength and still expose only the standard SMT footprint pattern. The overlap of the footprint by the coverlay should be a minimum of .010 inch or greater when space permits.

STRAIN RELIEF AND TEAR RESTRAINT METHODS

If the flexible circuit is going to be exercised excessively at one or more points, a strain relief or stiffener may be required. For example, an additional layer of the base material laminated to the stress-prone area

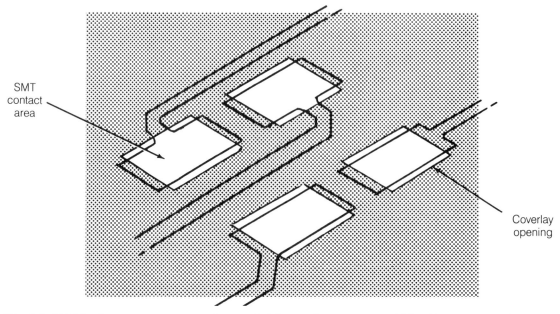

SMT
contact
area

Coverlay
opening

Fig. 2-5. SMD land pattern geometry must be extended for coverlay entrapment. This will ensure soldering and re-soldering of a component will not delaminate the copper from its base material.

will strengthen a connector. The thickness of the additional material can be .010 inch, if flexibility is needed or, as illustrated in FIG. 2-7, rigid materials can be added for maximum strength in an area.

Another zone requiring the addition of a thickness or rigid backing to the flex circuit is the SMT component mounting area. The small rectangular SMT patterns will sometimes promote a tear in the base material after the components have been mounted and soldered. This damage can be caused by stiffness of the soldered contacts and the weight of components during handling of the unsupported assembly.

SMT COMPONENTS NEAR A FOLD LINE

Mounting SMT components near a fold line should be avoided to reduce the tear factor. A clearance of .100 inch or greater from the component footprint pattern to the fold line will reduce stress between the base material and the copper foil. An extended layer of base material (FIG. 2-8) between the rigid backed portion of the SMT component area to the non-backed flex circuit will also prevent damage at a fold line. The fabricator should be consulted during the design phase of the product to determine the most practical solution to the problem. Lamination of

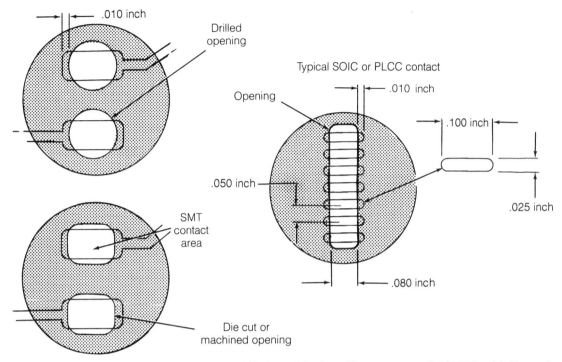

Fig. 2-6. Openings in coverlay materials can be drilled, machined, or die-cut to expose the SMT land patterns. In each configuration, a .010-inch coverlay overlap of the land pattern is provided to strengthen the solder area.

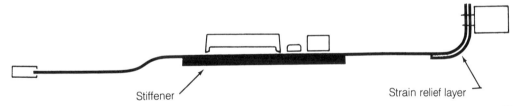

Fig. 2-7. Flexible and rigid materials can be laminated to specific areas of the flexible circuit to add durability or stiffness.

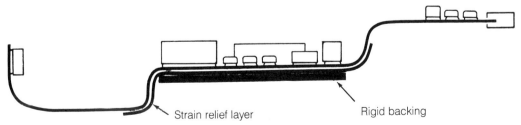

Fig. 2-8. An extension of a flexible strain relief layer laminated between the circuit and a rigid backing will reduce stress when a fold or bend of the circuit is near the edge of the rigid backing.

these tear restraints and strain relief layers is performed during the final stage of the flex circuit fabrication.

INSIDE CORNER SUPPORT

Tearing in the inside corners of a right angle bend in the flexible circuit is common. To correct this situation, retain a small pattern of the copper foil under the coverlay. This pattern will be inside the finished edge of the part and will follow the radii specified. Examples are shown in FIG. 2-9.

The same technique can be adapted to strengthen slots or narrow openings in a parallel appendage of the circuit. Widening the circuit trace paths in the corner or inside edge of the direction change can reduce tear stress as well. Keep in mind one point: avoid sharp corners on shapes added to the circuit. A radius at the end of patterns is preferred for all geometrical shapes.

PANELIZATION FOR VOLUME PRODUCTION

The smaller flex circuit assembly with components attached to the surface is difficult to process. Even when using special holding fixtures, the circuit material must be secured flat for solder paste application, component placement, and reflow. The fixtures to process these assemblies in volume can be expensive and labor intensive because each flex print unit must be located in position and secured to the fixture by hand. The circuit images are usually fabricated in panels and excised from the larger sheet as a final procedure. The assembly process is made easier

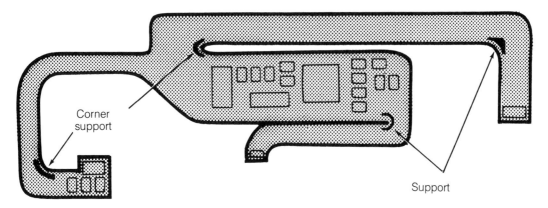

Corner support

Support

Fig. 2-9. Retaining small patterns of the etched copper at corner locations will furnish a barrier against tear points.

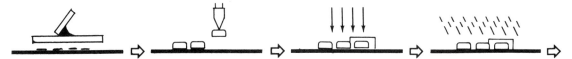

Fig. 2-10. The SMT assembly process for the flexible substrate starts with the application of a solder paste on the device land patterns, placement of the device into the solder, an exposure to heat adequate to melt or reflow the solder, followed by a cleaning cycle to remove the flux and other residue from the finished circuit.

by postponing separation of the individual circuits from the panel until all assembly and cleaning procedures are completed. The panel size would be reduced from the 12×24 inch sheet used to fabricate the circuit to a more manageable 12×12 inch or smaller multi-imaged set. A typical SMT assembly sequence for a flexible circuit would follow the one shown in FIG. 2-10.

The flexible circuits are then separated from the panel with the tooling, as originally planned in the fabrication of the circuit. An optional configuration would retain the pre-blanked units in the panel format with small tab connections. The die would pierce the base material in the finished shape except for a number of tab retaining points. These can be cut or punched as a post-process excising operation.

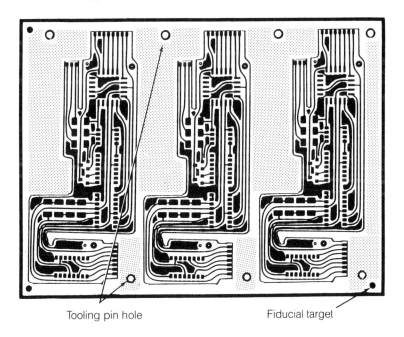

Tooling pin hole   Fiducial target

Fig. 2-11. Retaining the flexible circuits in a panel format, with fiducial targets, will assist in each phase of the automated assembly process.

FIDUCIAL (OPTICAL) TARGETS
FOR SMT PROCESS AUTOMATION

Vision systems for in-process handling of SMT assemblies have become commonplace in the sophisticated factory. High-resolution video cameras are integrated with the manufacturing systems to improve speed, accuracy, and to reduce labor. The camera can scan an entire assembly or specific areas or patterns. Target patterns must be provided on the substrate in order to use these advanced tools for processing SMT components on the flexible circuit.

Equipment manufacturers have recommended several shapes for the optical fiducial target. The pattern most acceptable for vision-equipped systems is a .060-inch diameter or solid-shaped diamond pattern etched into the copper material. This image must remain clear of other conductor trace images and coverlay coating. The clearance around the target should not be less than .060 inch to ensure recognition by the vision system. FIGURE 2-11 represents a panel array of several assemblies with all the features required for automated processing.

Fiducial target requirements for flexible and rigid circuits are defined in greater detail in chapters 8 and 9.

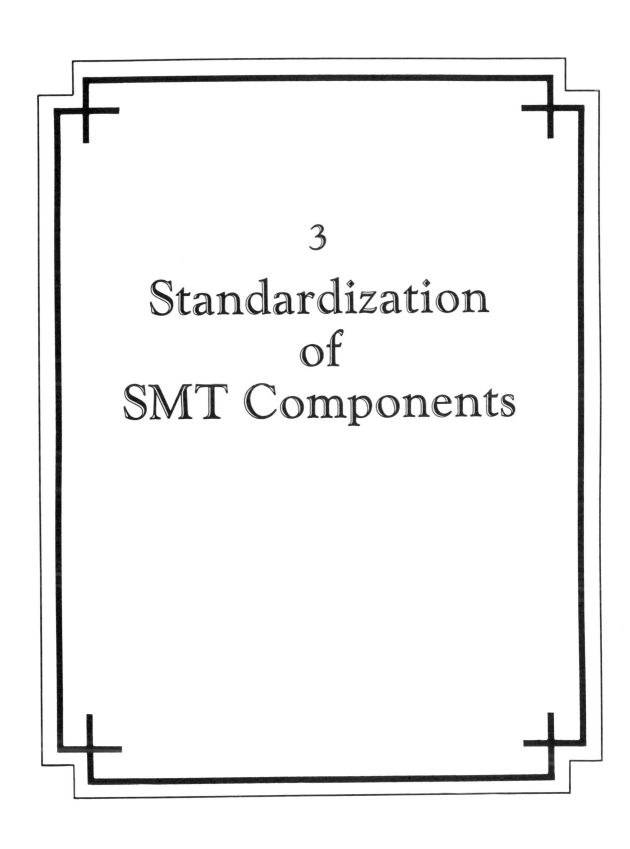

3
Standardization
of
SMT Components

BECAUSE OF THE NEED TO DEVELOP COMPONENT standards, organizations in the United States and worldwide have formed standards committees.

In the United States, organizations such as JEDEC, EIA, IPC and SMTA have played a valuable role in communicating the importance of standardization to component manufacturers.

In Japan, the counterpart to the U.S. JEDEC committee is the EIAJ. While the Japanese use the metric system, they maintain the .050-inch lead spacing on many IC components in order to be compatible with the American market.

Significant mechanical differences remain on Japanese products available with .050-inch lead spacing. For example, the distance between rows on the contacts of the EIAJ-SOP is not the same as the JEDEC SOIC. When specifying IC devices from multiple sources, it may not be possible to mount both Asian and domestic sizes on the footprint or pad geometry recommended by only one component manufacturer. Alternative ways to handle the variable size problem are discussed in chapter 4.

Component selection plays a major role in the design of SMT products and often presents unseen pitfalls to the new designer. The selection of a standard surface-mounted equivalent to a comparable leaded type device is a critical step which requires more than a casual look through component directories and data sheets.

Errors occur easily during the selection phase because options and styles of surface-mounted component packages may be unfamiliar to the component engineer and designer. Examples of a few common discrete component types are shown in FIG. 3-1.

While one surface-mounted device might look the same as the next, there are subtle differences in package types. For example, the same IC function might be available in two or more package styles from a single source.

Every year component manufacturers have increased the number of devices offered, but even today, component types supplied with conventional leads are not available or even considered cost effective for SMT. Continued demand by users will tip the scale in the favor of SMT and the industry will continue to see an increase in supply and a decrease in cost of surface-mounted devices.

Although standards have been well established for most surface-mounted devices, component engineers, designers, and purchasing personnel must work together to ensure the correct component is designed into and purchased for the product. It should be emphasized that the success of an SMT project is directly linked to the quality of communication between all members of the development team.

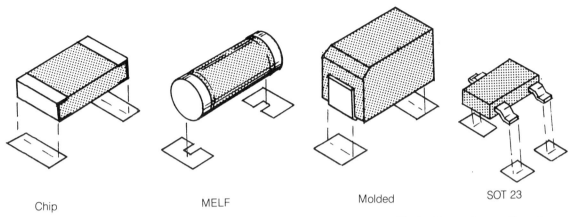

Chip   MELF   Molded   SOT 23

Fig. 3-1. Discrete components are available in standard EIA packaging that will adapt to most assembly systems.

COMPONENT PACKAGING OPTIONS

SMT components are supplied to the end user in one of three configurations: bulk, tube magazine, and tape-and-reel. Tape-and-reel packaging is preferable, depending on the component type, for medium- to high-volume production. With low-volume or prototype production, the tube magazine is recommended. Both containers are clearly marked for efficient material control and can adapt easily to the automated SMT placement equipment. Bulk packaging of chip type components is less desirable because the part must be handled with special feeders for the assembly equipment.

Chip resistors and capacitors are supplied from most sources on 8mm and 10mm tape-and-reel packaging. Each reel holds up to 4000 or 5000 chip parts. Component identification for material control is vital. The manufacturer's code number on the container is often the only way to verify contents, since not all chip component values are identified on the body surface.

The SOT (small outline diode, and transistor), like resistors and capacitors, are furnished in tape-and-reel. The orientation of the component in the cavity of the tape-and-reel must be specified by the user. The direction shown in FIG. 3-2, Option 1, is the most common and preferred standard, while Option 2 is considered special.

COMPONENT SELECTION

Choosing components for the surface-mounted assembly requires careful research and good judgment. When possible, select components

SOT 23 diode and transistor

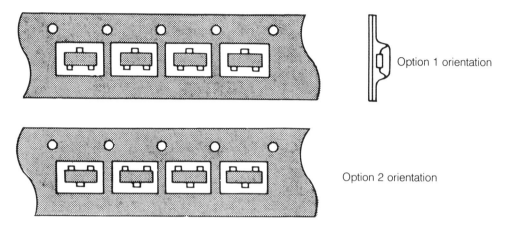

Option 1 orientation

Option 2 orientation

Fig. 3-2. Specify standard orientation for tape-and-reel packaged devices. Optional orientation is provided for older assembly systems that do not have full rotational placement capability.

that are uniform in type and size. An example might be resistors and capacitors that are available in the same dimensional shape. By using standard sizes, components will be available from several manufacturers at favorable prices. The mechanical elements of surface-mounted devices that adhere to the general industry guidelines are discussed on the following pages.

PASSIVE DISCRETE DEVICES

Monolithic Capacitors

The selection process for monolithic capacitors is more than choosing the value and the dielectric. Choices include body size, end-cap termination, and material or plating. The value range spans more than a dozen sizes from one manufacturer to another. To reduce unnecessary inventory of seldom used sizes, the EIA has standardized five sizes that have been recommended for use in the industry as detailed in FIG. 3-3.

The 0805 is used in applications requiring either maximum miniaturization or very low capacitance. The most common capacitor size for values up to 0.10μF is the 1206. The 1206 size device is considered a worldwide standard. The 0.10μF and the 0.15μF capacitor is available in the 1206 body, but the 1210 offers a broader source selection to the designer. The narrower component (1206) will mount on the 1210 foot-

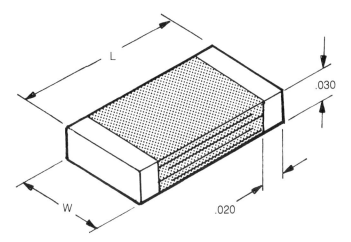

Fig. 3-3. Monolithic capacitor manufacturers supply devices in five standard sizes covering a wide value range.

print pattern. Higher voltage and capacitance values are available in the larger size with several dielectric options.

Note: Unlike the 0805 through 1812 devices, the 2225 size component is wider than it is long.

When specifying capacitors for reflow- and wave-solder processing, it is important to select the proper end-cap termination material. The most compatible end-cap termination plating considered standard is referred to as nickel barrier/solder type.

Molded Tantalum Capacitors

Manufacturers around the world have approved a set of standards for surface-mounted tantalum capacitors with a uniform molded body. The basis for these standards has been influenced by increased demand and the need for multiple sources.

There are many styles and sizes presently on the market, each with excellent mechanical and electrical characteristics. See FIG. 3-4.

Accepted levels of electrical and mechanical performance are required with each configuration, and at the same time, they must be compatible with the automated placement equipment.

Standards recommended by the EIA include two families of capacitors, each covering a specific value and performance range. The current standard range, shown in TABLE 3-1, includes values from $0.10\mu F$ through $68.0\mu F$. The extended range, TABLE 3-2, provides capacitance values through $330\mu F$.

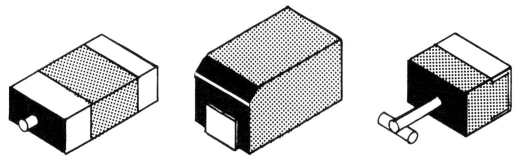

Fig. 3-4. The uniform molded body for the tantalum capacitor is preferred for assembly automation.

Table 3-1. Standard Range Tantalum Chip.

CASE SIZE	Max Capacitance (µF)/Voltage/Case Size Voltage							
	4	6	10	16	20	25	35	50
A	3.3	2.2	1.5	1.0	.68	.47	.33	.10
B	10.0	6.8	4.7	3.3	2.2	1.5	1.0	.33
C	33.0	15.0	10.0	6.8	4.7	3.3	2.2	1.0
D	68.0	47.0	33.0	22.0	15.0	10.0	4.7	2.2

Table 3-2. Extended Range Tantalum Chip.

CASE SIZE	Max Capacitance (µF)/Voltage/Case Size Voltage							
	4	6	10	16	20	25	35	50
A	10.0	6.8	4.7	3.3	2.2	1.5	1.0	.33
B	15.0	15.0	6.8	4.7	3.3	2.2	1.5	.68
C	100.0	68.0	47.0	33.0	22.0	15.0	10.0	3.3
D	330.0	150.0	68.0	47.0	47.0	33.0	22.0	6.8

Resistors for SMT

Surface-mounted resistors are available in all values from several sources. As with capacitors, some sizes are more common than others, and higher wattage components will have a limited value selection. The 1206 size, and greater, is available with values clearly marked on the surface for easy inspection.

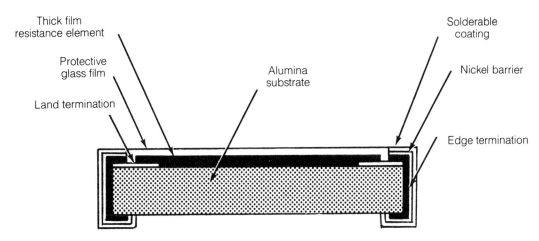

Fig. 3-5. Chip resistors are furnished in several physical shapes and wattage ratings. The rectangular 1206 size ceramic based device is considered a standard for ⅛-watt applications.

Chip resistors have a thick film element on an alumina substrate. The end-cap terminations have a tin-lead solder plating over a nickel barrier for compatibility with SMT assembly and solder systems. The resistor element is insulated with a resin coating. See FIG. 3-5.

The most common chip size in use today is the 1206. This size is usually rated at ⅛ watt by most manufacturers. It is often used for ¼ watt applications with specific temperature range limits. The standard resistance decade is shown in TABLE 3-1.

Other sizes are available as well: the 0805 for miniature applications with ¹⁄₁₀ watt rating and the 1210 device for ¼ watt applications. One-half watt and 1 watt resistors are available in larger sizes, but value selection is limited.

Resistors are also supplied in the MELF configuration, the tubular-shaped device with length and width dimensions similar to ceramic chip components. The MELF is popular for wave-solder applications when the component is held to the solder side of the substrate with epoxy. Before selecting the tubular shaped body, verify the compatibility of the component with the assembly equipment to be used.

Resistor Networks

Resistor networks are available in many surface-mounted configurations. Choose the component style that can be handled with automatic

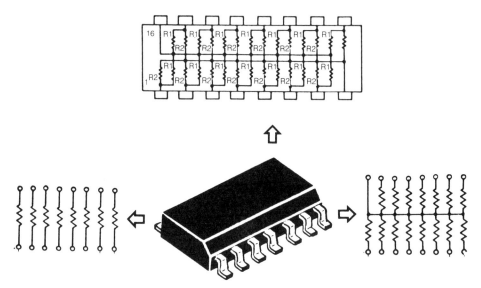

Fig. 3-6. Surface-mounted resistor networks are available in many standard values. Custom networks are ordered directly from the manufacturer.

placement equipment. When possible, find a style that is mechanically the same from two or more sources. See FIG. 3-6.

Networks are available from several major suppliers in SO-14 and SO-16 body types. Not all manufacturers conform to the JEDEC registered standard SO-14 and SO-16 with a .250 inch overall width.

Be aware of subtle differences when choosing component sources. When an incorrect substitution of components is made in purchasing, the result will be unwanted delays when the product goes to the assembly line.

Potentiometers for SMT

There is a multitude of single-turn cermet element potentiometers (pots) on the market for PC boards requiring a variable resistor. As the sources are researched, the designer will find that the parts differ from one manufacturer to another. Attempt to choose parts available from more than one source or select a very reliable single source.

Parts are available in open and sealed body types. It is important to choose one that is packaged for automated assembly equipment, as well as being usable in all types of solder-reflow processes.

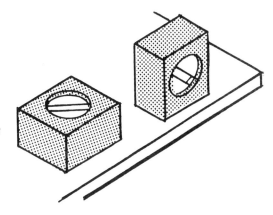

*Fig. 3-7. Potentiometer
standardization is not as
established as other component
types. Choose devices that can
be supplied from more than one
source when possible.*

The devices shown in FIG. 3-7 are designed for flat- and right-angle mounting. Many manufacturers offer these devices with contacts on .050-inch spacing, with open or closed body construction.

When mounting pots, a small amount of adhesive may be added to enhance the mechanical integrity of a part that requires frequent adjustments.

ACTIVE DEVICES

Transistors and Diodes for SMT

The most common case type used for Bipolar and Field Effect Transistors (FETs) are SOT-23, SOT-24 and SOT-89. The SOT-23 and SOT-24 are smaller, with three or four leads respectively. Details are shown in FIG. 3-8.

The pin assignments of the SOT transistor contacts are not the same as the TO-5 or TO-92 lead-type devices. In addition, the internal wire bonding may vary or may be optional with some manufacturers. The detail in FIG. 3-9 is typical for common transistors, but FET pin assignment varies, even with the same manufacturer.

Study the pin assignment specifications carefully before designing the PC board. An error at this stage of the project will have predictably negative results.

Diodes and Rectifiers

Most general-purpose diodes are available in the SOT-23, as well as MELF. Pin assignment of diodes to the three-lead component is optional

Standardization of SMT Components

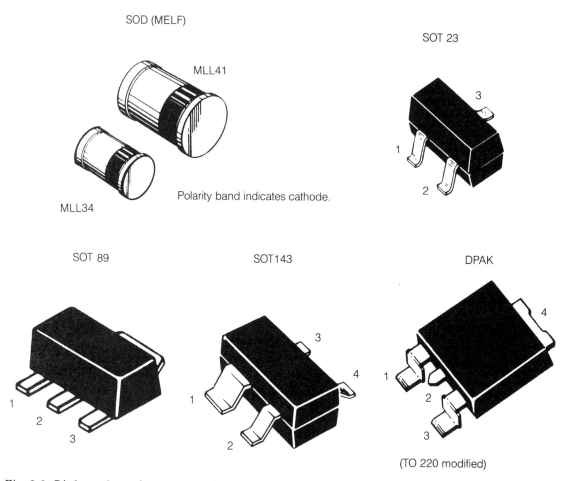

SOD (MELF)

MLL41

MLL34

Polarity band indicates cathode.

SOT 23

3

1

2

SOT 89

1

2

3

SOT143

3

4

1

2

DPAK

4

1

2

3

(TO 220 modified)

Fig. 3-8. Diodes and transistors are supplied in several standard physical configurations. Each is designed for a specific operational or environmental requirement.

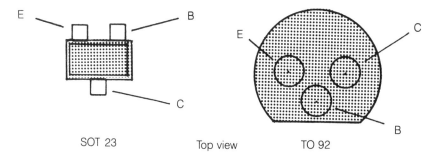

E   B

C

SOT 23   Top view

E   C

B

TO 92

Fig. 3-9. The internal bonding of the transistor die is not the same between the SOT-23 and its common leaded type TO-92 counterpart.

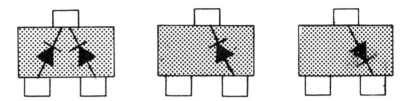

Fig. 3-10. General purpose diodes are available in the SOT-23 package with optional internal orientation.

from many suppliers, but to encourage standardization in the SOT-23, select the more common configuration shown in FIG. 3-10.

Dual diodes are also available in the SOT-23 with internal wire bonding of the die offered in several directional options. To reduce space on the assembly, as shown in FIG. 3-11 and FIG. 3-12, use two dual-diode components to create a bridge network.

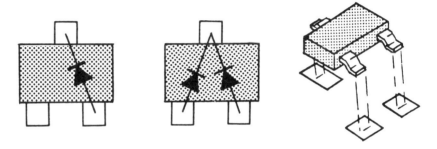

Fig. 3-11. Dual diodes and LEDs are supplied from several major manufacturers. Dual LED components are also available in two colors.

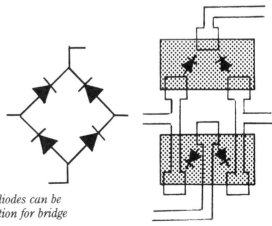

Fig. 3-12. Direction of dual diodes can be furnished with opposing direction for bridge network applications.

Standardization of SMT Components

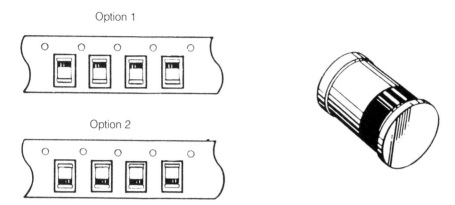

Option 1

Option 2

Fig. 3-13. The MELF package is the most common physical shape for general purpose diode and rectifier devices.

Transistor and diode arrays are offered in SOIC and PLCC type housings, but multiple sources are not always interchangeable. Choosing JEDEC packages will ensure standard sizes and compatibility with the circuit board design.

The MELF component style, which is available in the standard tape-and-reel packaging is often used for single diodes and rectifiers. Size will vary with AMP rating, type, and manufacturer. See FIG. 3-13. Further details on contact patterns and mounting methods are furnished in chapter 4.

ICs for SMT

Surface-mounted ICs are available in several body and lead frame styles: Small Outline (SO); Plastic Lead Chip Carrier (PLCC); and Quad Flat Pack (QFP). While manufacturers do not conform to any standard configuration, standards have been set for the SO, PLCC and QFP component families. In all cases, these surface-mounted ICs will be smaller than their leaded through-hole counterparts.

The JEDEC registered IC component types available today have a .050- and .025-inch space between lead centers. Specific contact geometry for these components is recommended by the component manufacturers and IPC.

The SOIC device shown in FIG. 3-14 is offered by several IC manufacturers. Most of the domestic manufacturers' configurations meet the JEDEC registered specification.

The dimensions shown in FIG. 3-15 reflect the JEDEC standard.

Fig. 3-14. IC package styles have been developed to adapt to any complexity of silicon. The SO IC is most commonly used for standard logic and analog devices. The PLCC and QFP packages will accommodate more complex or application specific requirements.

Other domestic and Asian manufacturers occasionally vary from JEDEC standards. Even though the .050-inch contact spacing is the same as the JEDEC part, the body width and distance between contact rows is inconsistent from one source to another. Examples of this variation are the 26- and 28-lead Static Ram (SRAM) ICs. If the circuit board is designed using a domestic IC's specification sheet, but an Asian part is substituted on the assembly, the leads of the device will overlap the narrower JEDEC footprint pattern. A contact design that will provide for this multi-width situation is illustrated in chapter 4.

Small Outline—SOICs

The JEDEC registered Small Outline, or SOIC contact pattern is easily adaptable to the circuit board. The two parallel rows of contacts have the same pin assignment as the dual in-line through-hole IC they replace. For most logic devices, the spacing between contact row centers will be consistent.

Depending on the manufacturer, the same IC function can be housed in different style components for SMT. The designer needs to be aware of this, as the pinouts will vary. An example of the same device

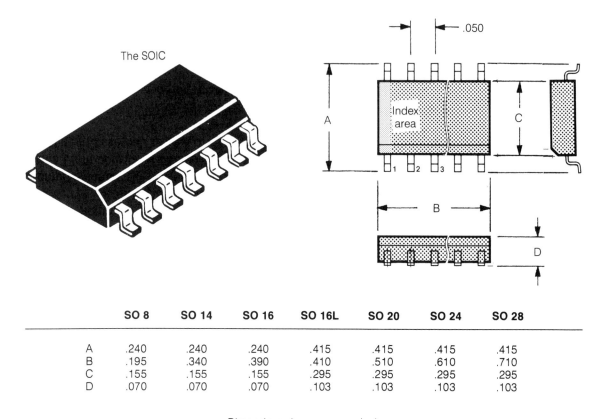

The SOIC

	SO 8	SO 14	SO 16	SO 16L	SO 20	SO 24	SO 28
A	.240	.240	.240	.415	.415	.415	.415
B	.195	.340	.390	.410	.510	.610	.710
C	.155	.155	.155	.295	.295	.295	.295
D	.070	.070	.070	.103	.103	.103	.103

Dimensions shown are nominal

Fig. 3-15. The JEDEC registered small outline IC package has a gull-wing lead shape on .050-inch center-to-center spacing and is standard.

function housed in the PLCC-20 and SO-14 is shown in FIG. 3-16.

In comparing these device types, the designer should note that the SOIC has the same pinout as the conventional DIP, while the PLCC-20 has a unique pinout with non-connected pins dispersed on each of the four sides.

Selection of the style of component is an important factor in planning the layout. Compare the SOIC footprint to the DIP that performs the same electrical function. First, the component profile above the surface of the board is much lower. Second, where component height is restricted, the SOIC offers a greater advantage over the DIP.

Some 16-pin devices require a wider lead frame to provide for the die size or the heat dissipation of the silicon. A PC board may have a mix

Typical 74LSO2 device

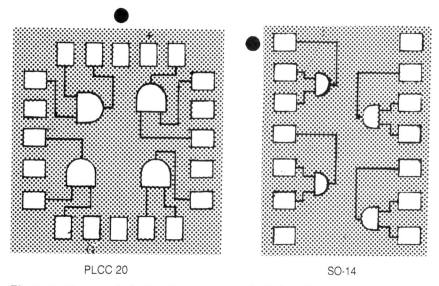

PLCC-20   SO-14

Fig. 3-16. The same device functions may be available in optional package styles. The component shape and pin assignment will be very different from one device type to another.

of the SO-16 and SO-16L (wide) devices. Always check the component manufacturer's specifications. SOICs greater than 16 leads will be in the wider lead frame.

The IC component buyer also has a choice in packaging. Tape-and-reel may be preferred by many companies with medium- to high-volume production. Tube magazine is recommended for lower volume prototype production, or when pretesting of each IC is required.

Plastic Lead Chip Carriers and Quad Packages

The most successful area of standardization has been in the plastic molded commercial grade devices. Several sizes are registered with JEDEC and comply to dimension and lead designs common to many domestic, European, and Asian component manufacturers.

The plastic lead chip carrier (PLCC) generally has an equal number of contacts on each of its four sides. The contact is formed in a J shape (the leads bent back under the package) to further decrease the surface area required for each device.

Another style of package similar to the PLCC is the C pack. The leads are on .050 inch or less spacing on four sides, but instead of the

J-bend, the leads are bent straight down to the board surface. This style requires a unique footprint for mounting. See FIG. 3-17.

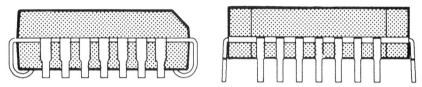

Fig. 3-17. The J-lead PLCC is considered a standard for the four-sided ICs up to 84 pins. Some manufacturers furnish a similar but nonstandard device in the I-lead or butt mounted configuration.

The PLCC is a commercial, high-volume component resembling the early ceramic leadless chip carrier (LCC). The PLCC has an excellent contact design and its symmetrical shape works well with most assembly equipment. The clearance under the PLCC allows for efficient cleaning after solder reflow.

The PLCC footprint is similar to the LCC; however, the J-lead design is better for reflow-solder processes and will adapt to a larger selection of substrate materials. The uniform sizes and J-lead design of the PLCC have been well received by the users of custom and semi-custom devices.

IC manufacturers have attempted to standardize package and lead design. The JEDEC committee has approved the PLCC in body configurations of 18, 22, 28, 32, 44, 52, 68, 84, 100, and 124 leads, as illustrated in FIG. 3-18.

Package sizes greater than 84 leads are difficult to place using robotic assembly. The machines are accurate enough, but the parts are bulky in size. In addition, the larger components cannot tolerate any deviation in the flatness of the substrate. IC suppliers have yield problems on the high lead-count devices due to excessive internal wire bond length. The delicate wire must span a greater distance from the silicon die to the lead frame. During the plastic molding operation, wire bonds are often washed away by the pressure of the molding process.

Some memory products are furnished in a PLCC-18 style device. Be aware that the PLCC-18 is supplied in two sizes and each has a rectangular shape with four leads on each end and five leads per side. The width of the IC is consistent, but the length will vary more than .060 inch. Check the manufacturer's specifications before board layout.

The 18-lead device is now available in two case sizes—18AA and 18AB. Details are shown in FIG. 3-19.

The PLCC

No. of leads	A	B
20	.395	.330
28	.495	.430
44	.695	.630
52	.795	.730
68	.995	.930
84	1.995	1.130
100	2.195	1.330
124	2.495	1.630

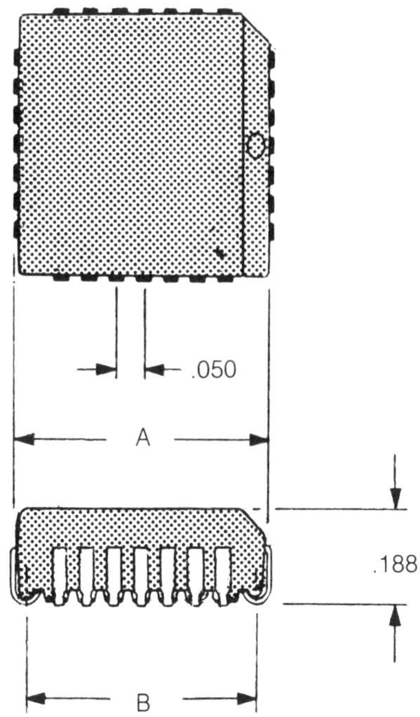

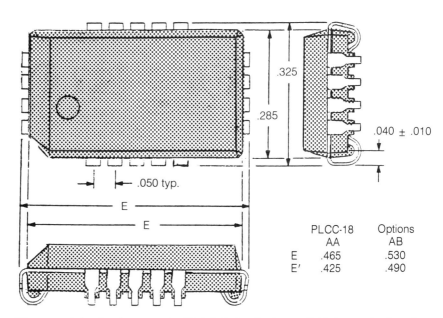

Fig. 3-18. The JEDEC registered PLCC IC package has .050-inch center-to-center lead spacing on four sides. This package is a common choice for custom and semi-custom gate array logic devices.

	PLCC-18 AA	Options AB
E	.465	.530
E'	.425	.490

Fig. 3-19. A family of rectangular PLCC devices is also available in standard configu rations typical of the 18 pin J-lead components illustrated.

47

The AB style allows the component supplier to furnish 18- and 22-lead devices with the same lead frame and plastic mold tooling.

Fine Pitch—High Lead-Count ICs

To better address the needs of the higher pin-count device, a miniature Plastic Lead Chip Carrier has become a practical and producible choice. These devices are generally referred to as a Quad Flat Pack (QFP) and will vary in size, lead spacing, and pin count.

As an example of one styling using .025-inch lead spacing, the detail in FIG. 3-20 illustrates a pre-formed, protected gull-wing lead configuration. The corner barriers, used in handling parts in carriers, protect the fragile leads during assembly operations.

Other Quad Flat Pack products are available as well, although not all are considered standard. From Asia, the 1mm and .8mm lead spacing is common for custom and semi-custom IC applications. In the U.S., at least one major IC manufacturer has a JEDEC registered high pin-count IC with .020-inch center-to-center lead spacing, or pitch.

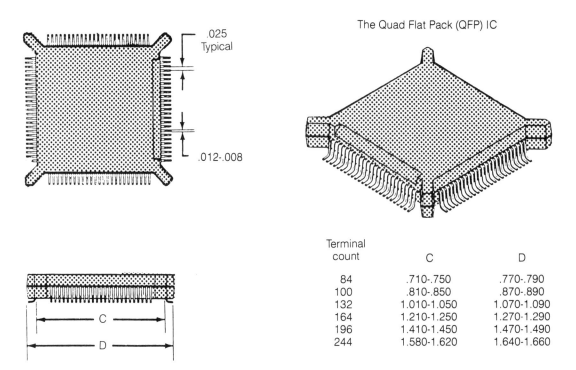

Terminal count	C	D
84	.710-.750	.770-.790
100	.810-.850	.870-.890
132	1.010-1.050	1.070-1.090
164	1.210-1.250	1.270-1.290
196	1.410-1.450	1.470-1.490
244	1.580-1.620	1.640-1.660

Fig. 3-20. JEDEC registered standard for the quad flat pack IC defines a gull-wing lead pattern with .025-inch center-to-center lead spacing.

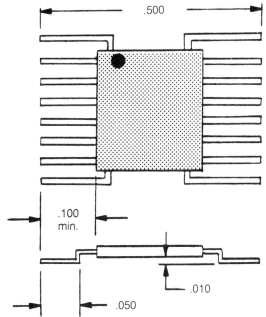

Ceramic flat pack (CFP) IC

Fig. 3-21. Ceramic surface-mounted devices are often furnished from the manufacturer in a straight unformed lead configuration. Tooling must be ordered to form and trim leads to the length desired.

These devices provide greater challenges to the assembly process by increasing the accuracy requirement on equipment and prompting a new generation of handling systems.

Ceramic Body ICs for High Reliability or Military Applications

Due to the various size differences in ceramic IC packages, a multiple source might not be possible for applications requiring a ceramic body. Most will have lead spacing of .025 or .050 inch with a lead length that is generally excessive and extends far beyond the component body. Therefore, leads on ceramic flat packs must be modified before mounting the IC to the substrate surface, as shown in FIG. 3-21.

4

Land Pattern Development for SMT

THE INSTITUTE FOR INTERCONNECTING AND PACKAG-
ing Electronic Circuit (IPC) Surface Mount Land Pattern
Task Group met in 1985 to initiate industry standards for attaching surface-mounted components to conventional printed circuit (PC)
boards. Members represented a cross section of both user and supplier
companies from various regions of the United States, Canada, Europe,
and Asia.

Initially, the task group focused on footprint pattern guidelines for
device types having the widest use in commercial applications. Over the
two years of data accumulation, the footprint study was expanded further than originally planned in order to address other component types
introduced during that period of time.

Most SMT components complied with the general mechanical limits
established through JEDEC registration. Today, a component style can
be registered, comply with a specific configuration, and still be quite different from one with the same function supplied by another manufacturer. The final publication for IPC-SM-782 was released in April 1987.
The footprint geometry recommended for the majority of SMT component types is discussed in this chapter.

CONTACT (FOOTPRINT) LAND PATTERN DESIGN

Contact patterns for surface-mounted devices vary due to package
shape, lead spacing and contact type.

In this chapter, the process-proven footprint patterns for each of the
component types presently used are illustrated and guidelines presented
for creating suitable patterns for future products. Footprint geometry
and spacing follow the recommendations and proposals furnished by
SMTA, IPC, EIA and leading component manufacturers.

Subtle changes in contact shapes for SMDs have evolved through
the years as the result of improved processes and refined component
quality. The contact geometry shown is primarily for the reflow-solder
process. Variations for wave-solder applications are also furnished.

DISCRETE COMPONENT CONTACT DESIGN

While resistors and capacitors are available in an array of sizes, the
designer should establish a uniform size for general use. A standard
footprint pattern gives assembly personnel greater control over the
equipment and processes. The most acceptable resistor and capacitor is
the 1206 size, which is used for $\frac{1}{8}$-watt and $\frac{1}{4}$-watt applications. Most
mid-range capacitors are available from several sources in the 1206 con-

figuration which will further standardize pad geometry and uniformity of the assembly.

The industry has standardized the chip component family into five shapes. Assembly equipment manufacturers suggest the 1206 or the 1210 size capacitor as the preferred choice. These capacitors are larger and easier to handle than the 0805. For values under 330pf, the 0805 is currently supplied. For values between 100pf and 330pf, the designer has a choice of either the 0805 or 1206 size. Capacitor values between 300pf and .18μF are widely available from several sources in the 1206 or 1210 size capacitor. The selection of size should be made early in the design because footprint patterns of the 0805 and 1206 (or 1210) are very different. See FIG. 4-1 for details.

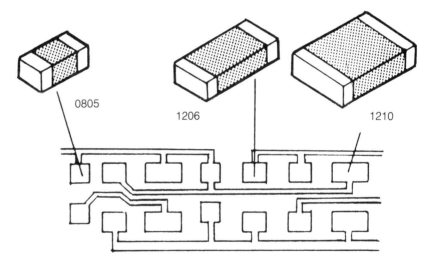

0805

1206

1210

Fig. 4-1. Value and dielectric requirements may force the use of several capacitor sizes on the same assembly. Changing from a lower to a higher value may require a modification of the substrate in order to provide the correct land pattern.

As previously noted in chapter 1, when placing plated through-hole pads on the artwork, do not allow the hole pad area to make direct contact with the component contact area. The clearance between the footprint pattern and the feedthrough pads will allow a desirable barrier of solder mask coating. This barrier will stop the migration of liquid solder during the reflow process. The conductor trace connecting the contact area with the feedthrough pads should be approximately .008 to .010 inch wide. See FIG. 4-2.

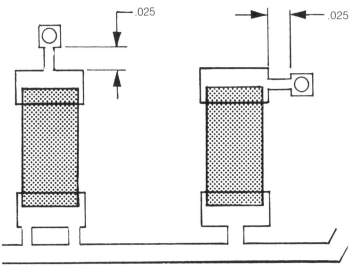

Fig. 4-2. Connect the SMT land patterns to a via hole and pad using a narrow circuit trace. Isolation of the contact by solder mask will stop the migration of solder in its liquid state.

Chip Component Pad Geometry

Pad geometry for chip or discrete components plays a prominent role in SMT assembly. Geometry is the most significant key to successfully controlling the reflow-solder process. Designers new to SMT must avoid the temptation to take short cuts in pad geometry. The examples shown in FIG. 4-3 describe a few of the Dos and Don'ts of pad geometry.

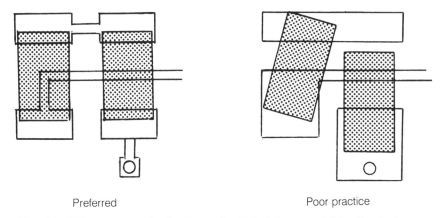

Preferred   Poor practice

Fig. 4-3. Chip component land patterns should be interconnected to other features with a narrow circuit, allowing equal surface tension through heat rise and cooling of the solder alloy during the reflow process.

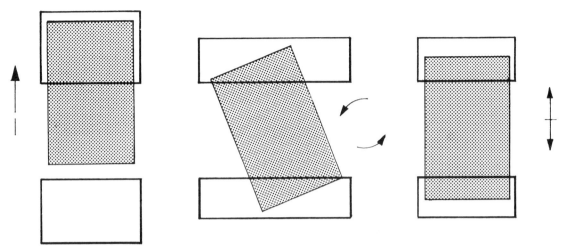

Fig. 4-4. The land pattern geometry is specifically designed to promote self-centering of each component while the assembly is exposed to the reflow-solder process.

Width of the pad for chip components is the first consideration for reflow-solder processing. If the pattern is too wide, the chip component could rotate, as illustrated in FIG. 4-4. Likewise, a footprint pattern that is too long will cause the chip component to float off the contact during cooling of the liquid solder material. The ideal pattern allows the liquid solder to cool evenly, centering the component equally on both pads.

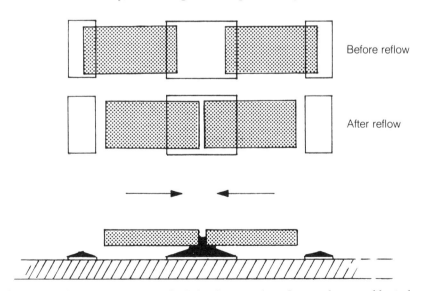

Before reflow

After reflow

Fig. 4-5. Avoid components on a single land pattern for reflow- and wave-solder technology.

Two chip components sharing one common contact between two components (FIG. 4-5) is not recommended. The solder on the larger pad will overwhelm the outer pads, drawing the components to the higher deposit of solder.

Component Spacing

During the reflow-soldering process, the solder is in a liquid form. While the solder is liquid, the components float on the high point of the contact pad. Problems occur when one component is spaced too closely to the next. One component might draw toward the other or slide off the center onto the adjacent pad. This problem is generally eliminated with adequate clearance between footprint patterns, as shown in FIG. 4-6.

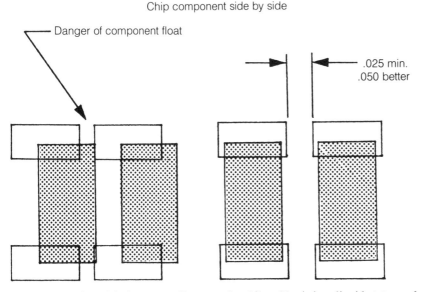

Chip component side by side

Danger of component float

.025 min.
.050 better

Fig. 4-6. Floating of devices onto adjacent pads while solder is in a liquid state can be avoided. Spacing between chip components should allow for a solder mask barrier.

To compensate for less than ideal pad geometry, it is necessary to use component mounting epoxy, normally reserved for wave-solder attachment. Adding epoxy to the reflow-solder process, as a band-aid for poor design, increases the cost of the assembly.

Other disadvantages of using epoxy in the reflow procedure are: 1) particles from the solder or flux can be trapped in the epoxy material, and 2) cleaning under the parts will be difficult, allowing metallic bridging to occur.

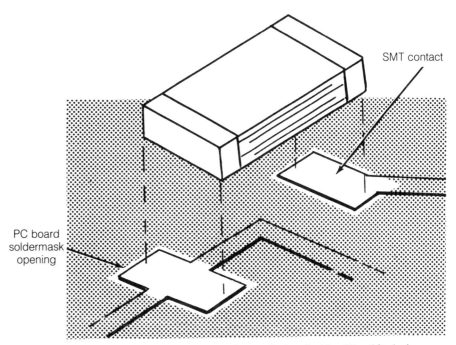

SMT contact

PC board
soldermask
opening

Fig. 4-7. The land pattern geometry must be equal at each side of the chip device contacts for the best solder process results.

The footprint pattern shown in FIG. 4-7 provides the most satisfactory results in the reflow process. This pattern also works well in wave solder when the placement of the component is accurate. Ideally, a finished solder connection encases the end-cap area of the component as shown in FIG. 4-8.

Fig. 4-8. Tin-lead alloy solder paste is momentarily converted to liquid to complete an electrical and mechanical attachment of the land pattern and device contacts.

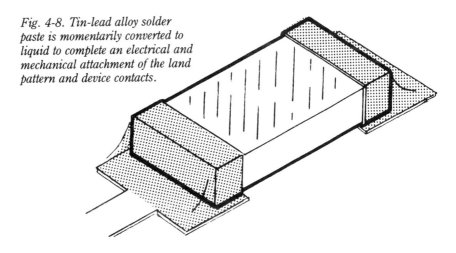

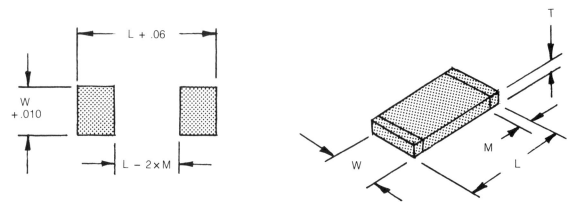

Fig. 4-9. Land pattern formula for chip resistors or capacitors has been developed for a stress resistant, reliable, and uniform solder termination.

BUILDING A CONTACT (FOOTPRINT) LIBRARY FOR SMT

As mentioned previously, resistors and capacitors are available in many sizes. Footprint patterns are illustrated for each of the five sizes recommended as standard. The contact design and spacing is typical of the formula shown in FIG. 4-9. This formula is adaptable to any chip style device.

The geometry illustrated in FIG. 4-10 provides a good electrical and mechanical interface, as well as promoting a self-centering reflow-solder process.

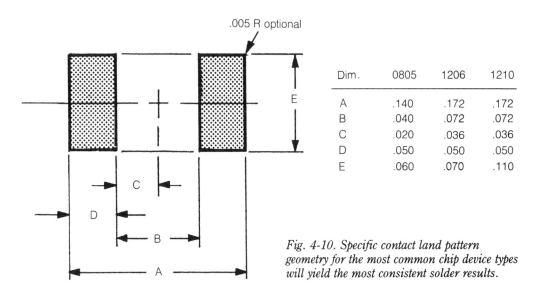

Dim.	0805	1206	1210
A	.140	.172	.172
B	.040	.072	.072
C	.020	.036	.036
D	.050	.050	.050
E	.060	.070	.110

Fig. 4-10. Specific contact land pattern geometry for the most common chip device types will yield the most consistent solder results.

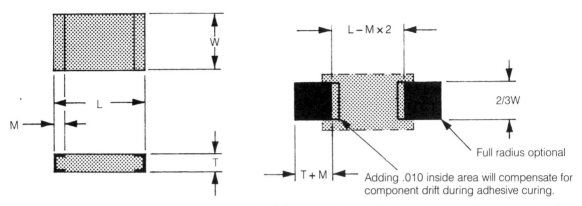

Fig. 4-11. Excessive solder buildup of epoxy attached devices can be avoided during wave solder by reducing the width of the land pattern.

Optional Wave-Solder Footprint Design

Wave-soldering chip components on the opposite side of leaded devices is popular, especially for PC boards with mixed technology. Often a company's first application of SMT employs this technique because equipment cost is not as significant and existing conventional wave-solder machines can be used. The full contact pattern works well in wave solder; however, some companies will choose a narrow contact area to control solder buildup (FIG. 4-11).

Clearances between lead pads of a component must be maintained. When component bodies are too close to an adjacent contact lead, bridging can occur during wave solder as shown in FIG. 4-12.

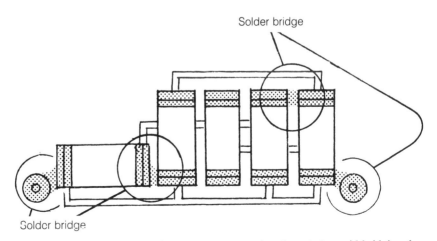

Fig. 4-12. Spacing between chip components must be adequate to avoid bridging during wave-solder processing.

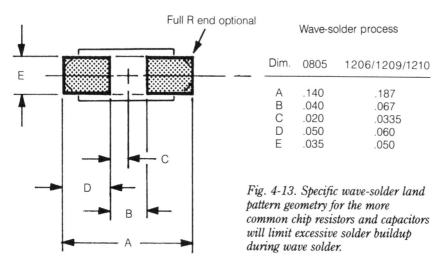

Dim.	0805	1206/1209/1210
A	.140	.187
B	.040	.067
C	.020	.0335
D	.050	.060
E	.035	.050

Fig. 4-13. Specific wave-solder land pattern geometry for the more common chip resistors and capacitors will limit excessive solder buildup during wave solder.

For those using wave solder to terminate chip components, it is acceptable to reduce the width of the contact pad geometry. Although the pattern shown previously works well in the wave-solder application, the narrow pattern minimizes excess solder on the miniature devices.

FIGURE 4-13 illustrates how pad geometry is designed to maintain a reliable solder connection while allowing for placement accuracy of assembly equipment and component shift during the epoxy cure.

MELF Component Contact Geometry

The MELF, a round shaped component, is dimensionally similar to the chip device and will adapt to the standard patterns detailed in FIG. 4-14.

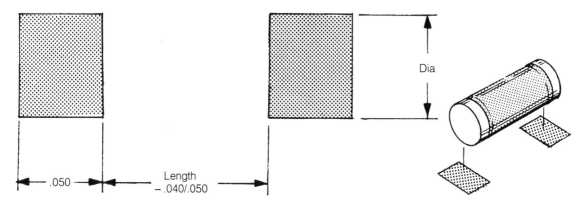

Fig. 4-14. Land pattern geometry recommended for MELF devices is suitable for wave- and reflow-solder processes.

When reflow-solder processes are used, the MELF device will occasionally roll off the center line. For that reason, adhesive epoxy, as used in wave solder, can be applied to retain the component on the component side. If the paste is the proper consistency and thickness, the epoxies can be eliminated when the MELF is reflow-soldered.

End-cap termination on MELF components is not always consistent from one manufacturer to another. Careful selection of the component supplier will help reduce the necessity of artwork changes or process variation.

SOT-23 Contact Geometry

The footprint pattern used for SOT-23 includes a pad for each of the three legs of the device. The SOT-23 patterns shown in FIG. 4-15 details one pattern recommended by the manufacturers and is best suited to wave-solder applications. The other pattern has an extra long footprint pattern on one side, which compensates for the imbalance of solder area during the reflow process. The solder does not cool equally on all three pads, and the side with two contact areas will dominate. Without this compensation, the component tends to lift away from the board surface on one side.

Many of the clearance rules used for chip components apply to SOT components. When arranging SOT components on the PC board, the designer must provide enough clearance to allow for the placement accuracy of the assembly equipment being used. The guidelines for contact area to feedthrough pad distance that are used for chip components are also valid for SOT devices, as shown in FIG. 4-16.

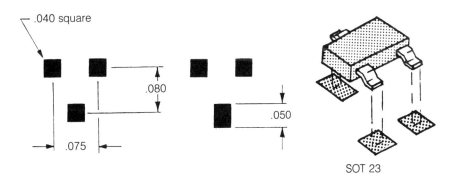

.040 square

.080

.050

.075

SOT 23

Fig. 4-15. The three point SOT-23 land pattern geometry of equal pad size will generally work for reflow- or wave solder. Extending the contact of the single offset pad will reduce the occurrence of lifting during reflow processing.

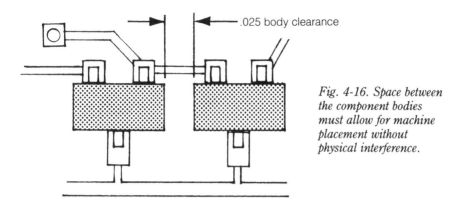

Fig. 4-16. Space between the component bodies must allow for machine placement without physical interference.

SOT-89 Contact Geometry

The SOT-89 package is used to accommodate the larger transistor die sizes and higher power operation of some devices. The large area of the center tab will aid in dissipating heat away from the component body to the surface of the board. The dimensions shown in FIG. 4-17 do not lend themselves to grid position, but they are recommended by the component manufacturer.

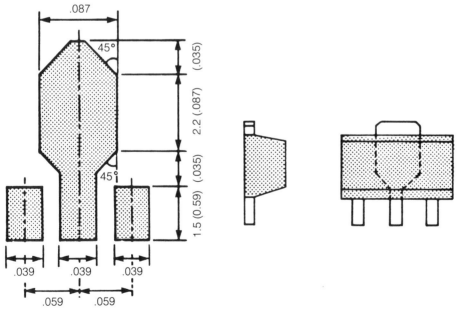

Fig. 4-17. Land pattern geometry for the SOT-89 power transistor provides a larger area of solder contact at the collector to improve heat transfer.

Small Outline IC Contact Geometry

The contact pattern for SOICs is designed to furnish a reliable electrical mechanical interface of the IC to the substrate and, at the same time, be *process friendly*. The term process friendly means a predictable, controlled method of attachment, using reflow-solder processes.

COMMERCIAL IC FOOTPRINT PLANNING

SOIC, PLCC, and Quad Lead Packaging

The Small Outline IC (SOIC) is assembled internally in the same way as the familiar DIP lead IC. The same die is attached to a smaller lead frame. The wire is bonded from the silicon die to the lead frame, then the plastic body is molded, encasing the entire assembly. The major difference between the SOIC and the DIP IC is that the SOIC will be mounted with solder to a footprint contact on the surface of the substrate instead of to through-holes.

The SO-8, SO-14 and SO-16 ICs use a smaller die size. The lead or pin assignment of the SOIC is usually the same as the larger DIP package. It is advisable to check the specifications carefully. Some SO-16 devices require a larger lead frame to accommodate the die size. This wide format, referred to as the SO-L or SO-W, extends to the SO-20, SO-24, and SO-28 as well. See FIG. 4-18.

Pad geometry can be adjusted to allow substitution on non-JEDEC SOICs. By lengthening the contact area, a wider-than-standard part can be mounted to the PC board with minimal impact on the assembly process. FIGURE 4-19 illustrates a typical solder connection to the PC board.

A pattern similar to the one shown in FIG. 4-20 may be used as a solution to the second source problem when the same function is available in the narrow package from Vendor "A", while Vendor "B" supplies only the wide package.

Note: This pad geometry takes more space on the board and is not a practical solution for all situations.

Plastic Lead Chip Carrier (PLCC) Contact Geometry

The PLCC footprint is similar to the leadless chip carrier (LCC), but the superior J-lead design is more suitable to reflow-solder processes and the broader selection of substrate material.

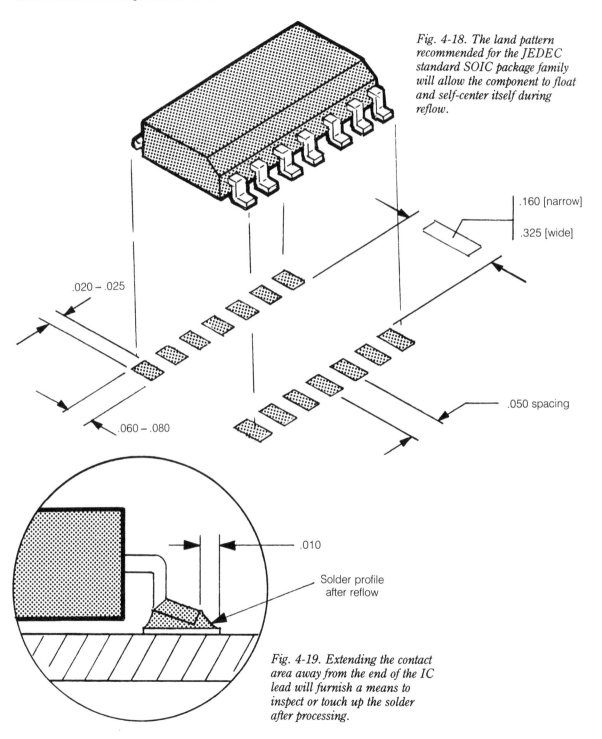

Fig. 4-18. The land pattern recommended for the JEDEC standard SOIC package family will allow the component to float and self-center itself during reflow.

.160 [narrow]

.325 [wide]

.020 – .025

.060 – .080

.050 spacing

.010

Solder profile after reflow

Fig. 4-19. Extending the contact area away from the end of the IC lead will furnish a means to inspect or touch up the solder after processing.

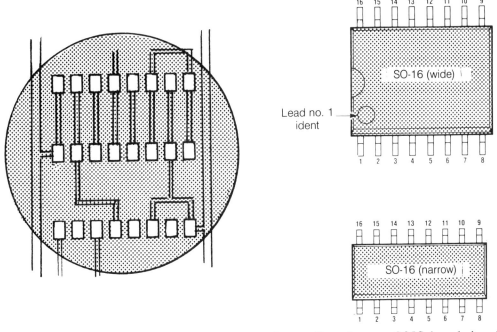

Fig. 4-20. On rare occasions, a land pattern must accommodate a wide and narrow SOIC for a device with the same function from two different sources.

The even number quad pattern provides consistent wire-bonding length and lead frame design. The PLCC devices are available in 20, 28, 44, 52, 68, and greater lead patterns with contacts on preferred .050-inch center-to-center spacing.

FIGURE 4-21 illustrates the subtle contact pattern offset which increases producibility and provides for inspection or rework.

The ideal PLCC footprint allows a good solder fillet toward the outside of the IC package. Extending the footprint inward only increases the chance of component drift during the reflow process. Extending the footprint's length further to the outside is acceptable for test-probe only if isolated by a narrow connecting trace. Test probe contact on the component side will reduce usable board area and component density. See FIG. 4-22.

The distance between contact groups is usually determined by the complexity of the circuit. When bussing common signal traces, a greater number of contacts of ICs can be interconnected on internal layers allow-

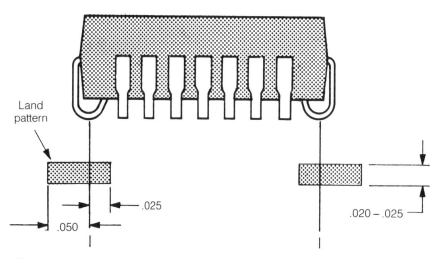

Land
pattern

.025

.050

.020 – .025

Fig. 4-21. The PLCC land pattern is offset from the center-line of the J-lead to provide for a uniform solder fillet and easy visual inspection.

ing higher density of components. In these cases, a minimum of .025 inch between footprint contact rows is possible, but for access and inspection of PLCC components, allow .150 inch or more space between adjacent component leads. (See FIG. 4-23.)

All footprint patterns for the PLCC should allow the designer the option of routing a .008-inch trace between contacts while maintaining a

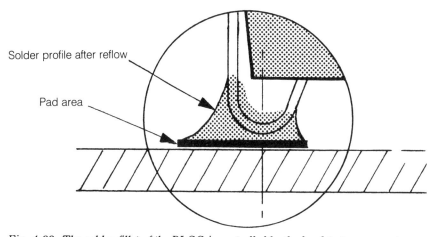

Solder profile after reflow

Pad area

Fig. 4-22. The solder fillet of the PLCC is controlled by the land pattern geometry.

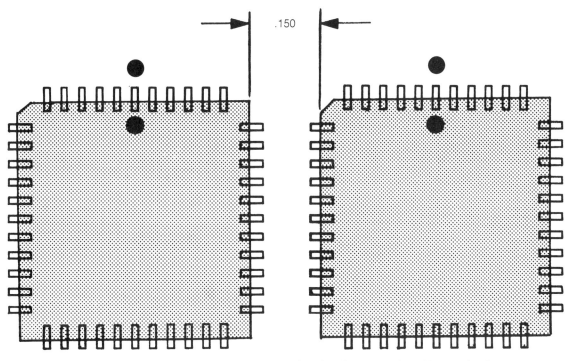

Fig. 4-23. Spacing between higher profile devices must provide visual access to the solder termination.

.008-inch air gap on both sides. The patterns shown in FIG. 4-24 accommodate the most common JEDEC PLCC devices.

The footprint for the PLCC will be separated from a via or feed-through hole by a narrow connecting trace. FIGURE 4-25 is typical of the contact and feedthrough pad interconnect on the PC board surface. The shaded area represents the solder mask coating over all conductor traces.

Quad Flat Pack ICs

Quad lead flat packs are one of the popular surface-mounted packages for custom and semi-custom ICs. Contact spacing will vary from one manufacturer to another and could range anywhere from .050 inch, .040 inch, 1mm, 0.8mm, down to .025 inch, .020 inch and less.

The Quad Flat Pack IC in FIG. 4-26 is more difficult to handle in volume production. The closely spaced leads are very fragile and require precise machine placement and a refined solder process.

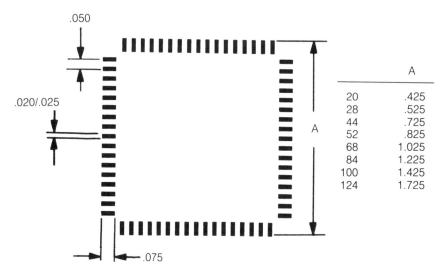

.050

.020/.025

.075

A

	A
20	.425
28	.525
44	.725
52	.825
68	1.025
84	1.225
100	1.425
124	1.725

Fig. 4-24. The land pattern family for square PLCC devices extends from 20 through 124 leads. PLCC lead count beyond 84 pins is generally avoided.

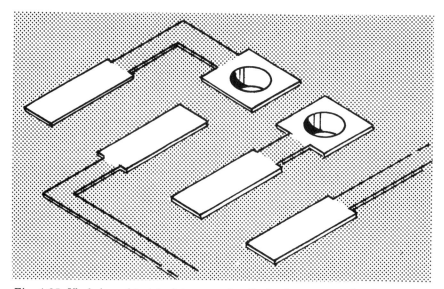

Fig. 4-25. Via hole and pad for interconnecting the land pattern to other layers of the substrate must be separated by a narrow circuit trace.

The footprint pattern will vary depending on lead spacing and length. The guidelines, shown in FIG. 4-27, are generally accepted as reflow-solder compatible. Components are often supplied in partitioned tray carriers with a protective recessed area for each device. Not all

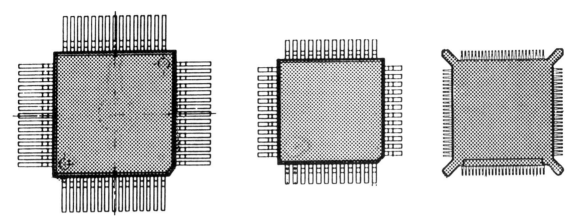

Fig. 4-26. QFP devices have a greater lead count in a smaller area than the PLCC family. Spacing between leads often varies from one manufacturer to another.

equipment adapts to this system and special fixtures must be developed. Components should be furnished with the leads pre-formed to mount directly onto the substrate surface. When the leads are not preformed, the company must specify bend and length desired when ordering devices.

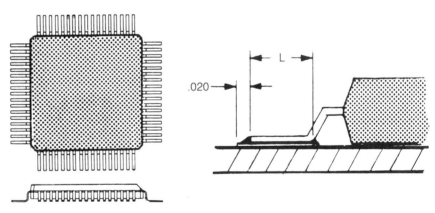

Fig. 4-27. QFP devices will require special equipment for placement. Furnish approximately .020 inch of contact area at the lead end for solder touchup and inspection.

CERAMIC IC DEVICES FOR SMT

For military or extreme environments, it is necessary to specify ICs in a ceramic package. As with commercial products, there are several configuration options:

- Leadless ceramic chip carrier (LCC),
- Ceramic quad with J-bend leads (LDCC),
- Ceramic flat pack (FP),
- Ceramic quad flat pack (QFP).

Ceramic packages as shown in FIG. 4-28 are packaged for lower volume applications, but can be handled by many assembly systems available. Leadless devices must be mounted to a substrate material more stable than the commercial substrate equivalent, because of the difference in thermal coefficient of expansion (TCE). This will cause failure of the solder connection with repeated temperature cycling of unlike materials.

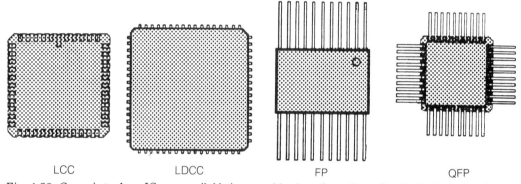

LCC   LDCC   FP   QFP

Fig. 4-28. Ceramic package ICs are available in several lead configurations. Specify device leads trimmed and preformed for direct mounting when possible.

Ceramic components (FP) with leads extending straight from the sides are usually modified before being attached to the substrate (FIG. 4-29).

The advantage of using the ceramic products with leads is the compatibility with a large variety of substrate types. The leads will flex sufficiently during the temperature variables, acting as shock absorbers for the solder bond, eliminating the TCE problems.

Caution: Pin #1 on the contact area of the ceramic LCC device is longer than the others for easy identification. If the mating contact on the substrate is also lengthened, it will act as a visual orientation guide, thus avoiding the danger of unwanted interconnection to feedthrough pads or traces exposed to the contact.

DIP AND SIP MODULE DESIGN

The Dual In-line Package (DIP) module shown in FIG. 4-30 is a method of taking advantage of high-speed robotic assembly technology

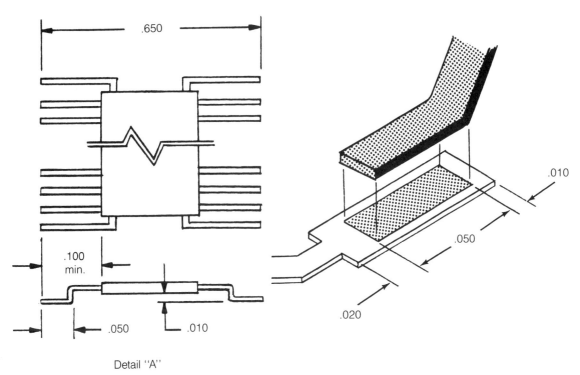

Detail "A"

Fig. 4-29. Minimum surface contact area is specified in DOD 2000 for surface-mounted ICs.

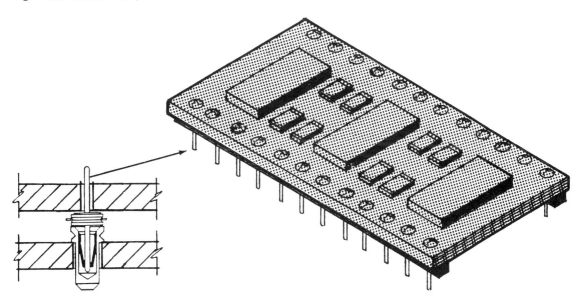

Fig. 4-30. Custom modules have several interface contact options. The DIP format shown will allow the module to have a very low profile (but can easily unplug form sockets furnished in the mating substrate.)

71

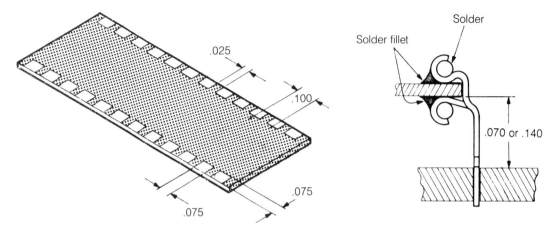

Fig. 4-31. Edge clip contacts are economical for direct soldering of DIP modules into a primary substrate. The most common lead spacing is .100-inch centers.

on a smaller, controlled scale. The size of the DIP module could conform to the limits of standard ICs or could take an outline adapted to individual needs. The designer is not limited to mounting components on one side of the module. It is a common practice in SMT, as well as in hybrid assemblies, to use both surfaces to increase the density of the module.

Dual in-line contact strips are the most economical method of terminating this module. The contact strips work well when the module is mounted into assemblies that are to be wave soldered. The footprint pattern shown in FIG. 4-31 is for reflow or dip soldering the contacts to the module. Pin and socket strips or headers should be used for more durable contact requirements. Typically, pin and socket connectors are used when a module in a socket needs to be added or replaced without special tools.

The edge-mount contacts are furnished in a continuous strip with a common breakaway bar to help retain alignment. Contacts can be supplied with solder preforms built into the contact to ensure an even flow and a strong connection. The contact strips are best suited for modules requiring a low profile.

The Single In-line Pin (SIP) module is popular for SMT. Ceramic hybrid modules often are converted to surface-mounted assemblies, thereby reducing substrate costs by 50 to 60 percent. It is common to partition the electronic functions into a module that can be used in many products. See FIG. 4-32. The SIP module is handled as a component, tested and easily soldered into larger PC board assemblies.

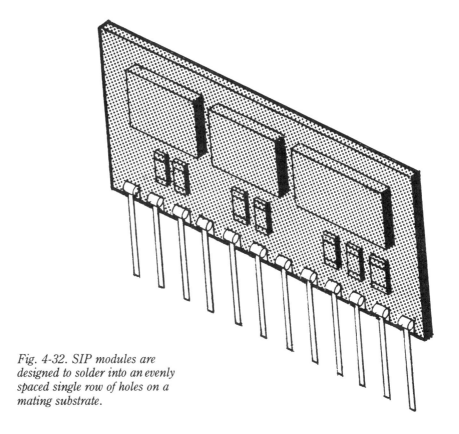

Fig. 4-32. SIP modules are designed to solder into an evenly spaced single row of holes on a mating substrate.

Contacts for SIP modules are also supplied plain or with solder to help the uniformity of the solder connection. SIP contacts can be mounted to the module prior to the solder-reflow process or added as a post-assembly procedure.

The footprint pattern shown in FIG. 4-33 provides excellent electrical and mechanical bonding characteristics after reflow-solder. When allowing for overall height of the SIP module, don't overlook the stand-off height built into the contact. This height is optional and must be specified by the user. It is important to study the manufacturer's specifications closely. Contact pins are designed to mount into hole patterns and sizes generally used for DIP ICs or other leaded devices.

STANDARD SMT MEMORY MODULES

The DRAM (Dynamic Random Access Memory) is one of the most common forms of on-board memory. Memory devices can be mounted

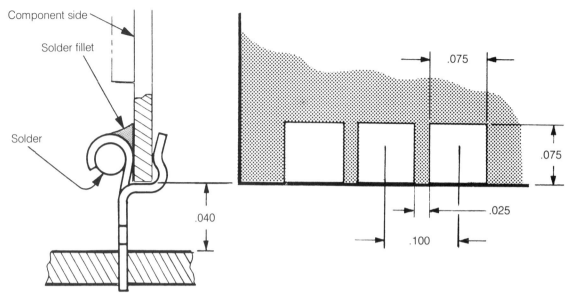

Fig. 4-33. The edge clip contact area is generally furnished with .100-inch center-to-center lead spacing.

directly to a PC board or furnished as an expansion or extension to the on-board memory. DIP style memory ICs have been added to conventional PTH boards using sockets. The miniature SMT ICs, however, are difficult and expensive to socket as individual components, but easy to add as a modular plug-in set.

The examples shown in FIG. 4-34 are typical of the JEDEC registered module designs furnished by several suppliers. This standardization has assisted distributors and users of memory in maintaining several approved sources.

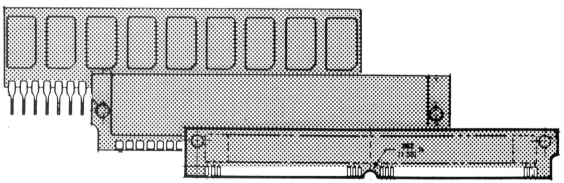

Fig. 4-34. SIMMs have several standard configurations. Solder mounted connectors have been developed to mate with the modules.

The SIMM (Single In-line Memory Module) will plug into a specially designed connector that will lock the module in position. Sockets are available from several sources and provide for vertical 90 degree mounting and angle mounting for low-profile applications.

The JEDEC registered SIP assembly is very similar to the SIMM except that extensions required for socket retention have been deleted and pins are soldered on the edge contacts. These pins will adapt the SMT memory module to a PTH-type PC board in the same way as the custom SIP module noted above.

SOJ CONTACT GEOMETRY AND SPACING

The SOJ package, although available for several applications, is used most extensively for 1 MEG and 4 MEG DRAMs. The footprint options shown in FIG. 4-35 illustrate a standard geometry and an optional shape that will accommodate the minimum spacing of the JEDEC memory modules. Both shapes are compatible with reflow-solder processes. Although the land area of the closely spaced ICs is reduced, adequate solder mass will be provided for a reliable connection. Both SOJ footprint pattern options will promote self-centering during the reflow process, also common to the SOIC.

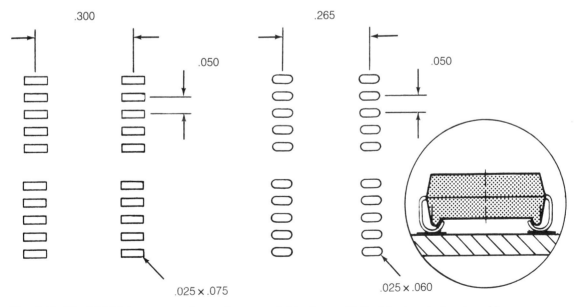

Fig. 4-35. The SOJ IC is a rectangular surface-mounted package with the J-lead contact on two sides.

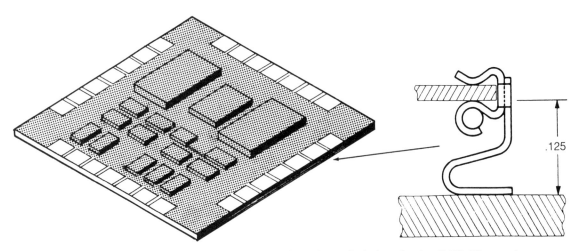

Fig. 4-36. Edge clip contacts are available for lead spacing of .050 inch for adapting LCC ICs or other custom modules to conventional substrate materials.

When the SOJ devices are mounted on the minimum clearance required for the SIMM or SIP assembly, trace routing on the component side is quite difficult. The alternate and pattern geometry will allow memory interconnection with minimum use of via or feedthrough holes.

QUAD MODULE DESIGN

Using the .050-inch space contact leads, the designer can develop an SO and quad module or adapt a ceramic LCC to a conventional Fiberglas PC board. Use the SMT carrier contact when possible and design the footprint pattern around one of the JEDEC standard arrangements. If the IC is then converted to a plastic molded device, the main PC board will not require modification.

Carrier contacts, shown in FIG. 4-36, are supplied on a .050-inch spacing and are ideally suited for miniature applications. The compliant contact design allows the mating of substrate materials and absorbs stress caused by different rates of expansion during thermal cycles.

Further complex interfacing between a module and the main PC board is possible with closer lead spacing. The chip carrier design is often used to provide a working model of a proposed custom gate array or for a product that is not presently available in a surface-mounted package from the component manufacturer. Using this carrier module concept may provide for the strategic, early entry of a product into a highly competitive marketplace.

5
Space Planning and Interface

SPACE STUDY FOR SMT

The space study, a preplanning procedure, will assist the designer in the organization of components and in the estimation of the density factor on the board (FIG. 5-1).

The primary goal of the pre-planning stage is the allowance of adequate space between components, and the clearance needed to assure an easy-to-manufacture assembly. As component density increases on the circuit board, some problems must be addressed: conductor width must be reduced; additional circuit layers will be added; and inspection, testing, and rework become more difficult.

Estimating Total Component Area

All footprint dimensions of the SMD must be included in the preliminary planning of the board (FIG. 5-2). The component body dimensions alone do not furnish sufficient accurate information to determine board space.

A percentage multiplier is added to the footprint area to allow for conductor traces and feedthrough or via hole pads. FIGURE 5-3 illustrates three density factors. As the density factor approaches a ratio of 1.5, the designer must select more circuit layers to provide signal interconnection or transfer a number of components to side two.

Interconnection of dissimilar component types will require additional planning. The SOICs shown in FIG. 5-4 require less surface area for routing than the PLCC. Mixing the two types on the board is common, but routing of traces between devices will be more difficult.

Choosing components that make maximum efficient use of space will best facilitate interconnection.

Optimal Component Placement

During the initial placement phase, the designer plans the optimum location of components for efficient interconnection of related devices. Interconnection of components with conductor paths on the same surface as the components can present a challenge to the designer. It is this interconnection factor that requires careful attention to the relationship, orientation, and placement of the SMDs as shown in FIG. 5-5.

The arrangement of components in a functional relationship to each other is only one consideration in the planning of the surface-mounted PC board. The illustration shown in FIG. 5-6 positions components for

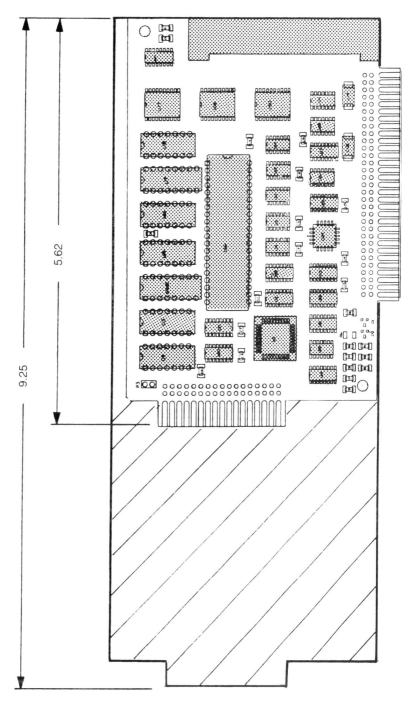

Fig. 5-1. The density of the surface-mounted assembly is often double the pin-through-hole equivalent of the same function.

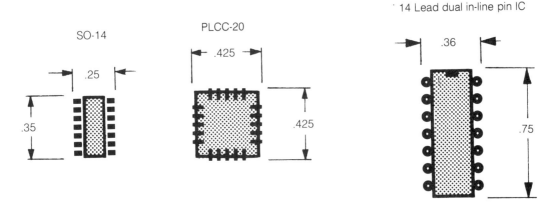

Fig. 5-2. *The Small Outline, Plastic Lead Chip Carrier and Dual In-line Pin IC packages illustrate the dramatic space reduction possible using SMT.*

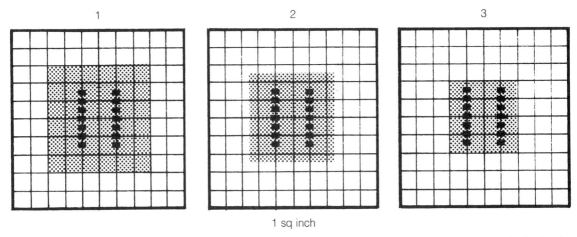

Fig. 5-3. *Density factor is measured by the space required to mount and interconnect the components. As the density factor compresses or space around each device is reduced, circuit layers must be added to accommodate circuit routing.*

the most direct interconnection possible. This direct coupling for analog or layout-sensitive circuits is not uncommon, but will subtly add to the cost of manufacturing the product.

Note: Regarding the direction of the SOT devices, ICs and other chip components—overall manufacturing time increases each time the component orientation changes.

SOT devices are furnished in both tape-and-reel form and tube magazine, with either right- or left-orientation. Even though assembly

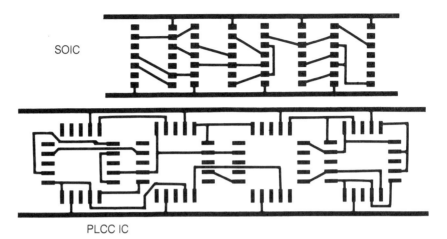

SOIC

PLCC IC

Fig. 5-4. The density factor is affected dramatically by the package type selected for the circuit. Logic components in the SO package will require less space for routing circuit traces when compared to the PLCC IC interconnection.

equipment will rotate each part before placement, cycle time is often slowed. The designer should plan the layout carefully to maintain a consistent component orientation.

PREFERRED ORIENTATION FOR SMT ASSEMBLY

Proper component orientation, as illustrated in FIG. 5-7, is vital to maintaining an efficient, cost-effective assembly. Both component orien-

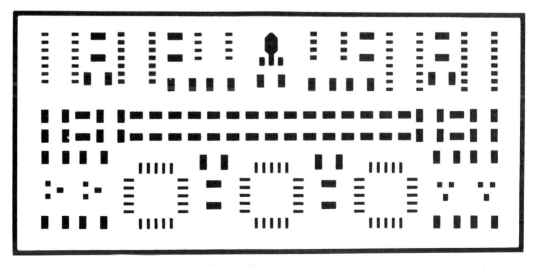

Fig. 5-5. Organization of the component layout will simplify the assembly process and reduce manufacturing cost.

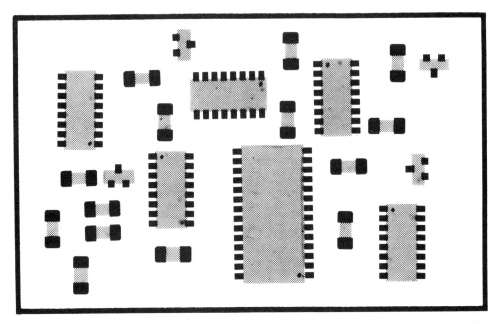

Fig. 5-6. Direct coupling of devices on the substrate is often necessary for analog or RF circuits. The difficulty and additional labor required for assembly is generally an accepted cost associated with these products.

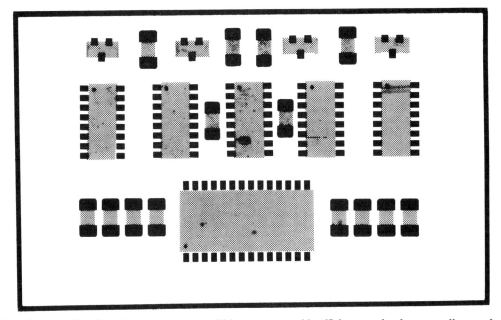

Fig. 5-7. Common direction of component parts will increase assembly efficiency and reduce overall manufacturing cost.

tation and signal paths are planned to take advantage of the board's surface area. IC placement is often determined by the relation to interface connectors. This, in turn, reduces excessive conductor crossover and allows shorter signal trace length. Related or interactive devices are grouped into functional clusters to make the most direct circuit trace interconnection possible.

SMT placement equipment allows for a 360-degree rotation of parts, giving the designer unlimited flexibility in placement orientation. Consistent orientation of components will expedite the assembly process, a cost-saving factor that must be considered at all times during the design phase of the project.

FINE PITCH, QFP, AND SOIC SPACE PLANNING

Fine pitch devices with center-to-center lead spacing of .025 and .020 inch, or less, may require a reduced volume of solder paste for reflow processing. Specific spacing between the contact area of the QFP ICs and other fine pitch devices must be maintained for solder paste application.

If a stepped or multilevel solder stencil is to be used, a minimum area of .125 inch must be provided between the overall contact patterns of adjacent components with wider lead spacing. The detail shown in FIG. 5-8 is an example of a group of SMT components including a typical fine pitch device. The QFP ICs may be mounted and reflow soldered at the same time as other surface-mounted devices or attached as a secondary, single station assembly process. The post-assembly operation may include a more precise placement system or a unique reflow-solder process. A detailed description of solder stencil fabrication and assembly options will be included in chapter 9.

USING BOTH SIDES OF THE SUBSTRATE

Unless all components chosen for the assembly are available in a surface-mounted configuration, mixing surface-mounted components with leaded devices will be unavoidable. One option open to the designer is the two-sided assembly. The majority of the active components are mounted on side one, while chip components are mounted on the opposite (wave solder) side, or side two.

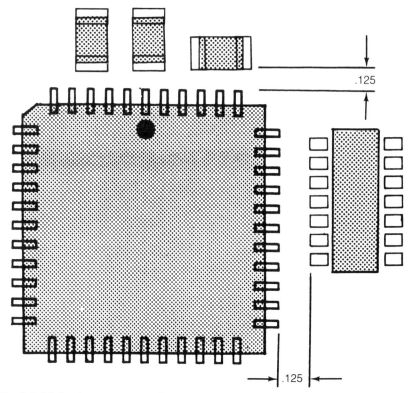

Fig. 5-8. Maintain a component clearance around the fine pitch QFP device to accommodate etch stepping of solder paste stencils.

When leaded (PTH) components are installed, the assembly is usually passed through a wave-solder process. By mounting the majority of the low-profile chip components on the wave-solder side, more surface area is reserved for trace interconnection on the component side. The low-profile chip components are generally well inside the finished length of the lead ends of PTH components. Resistors and capacitors are attached to the wave-solder side of the assembly by adhesive or by a UV cured epoxy developed specifically for the wave-solder process.

Locating surface-mounted ICs under leaded devices will maximize the area and volume of the PC board as shown in FIG. 5-9.

Note: Be aware of an increased occurrence of solder bridging between IC pins when wave soldered. If ample clearance is not reserved, touch-up will occasionally be required even with advanced dual wave-solder techniques.

Careful placement of surface-mounted chip components will take advantage of small spaces unused by bulky leaded counterparts. By plac-

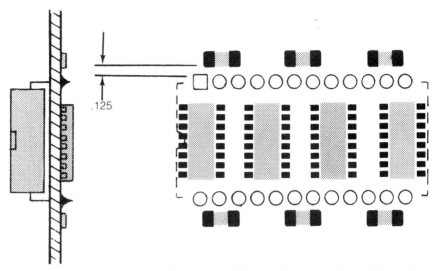

Fig. 5-9. Wave solder of leaded parts often follows reflow processing of the surface-mounted components. Allow spacing between the device leads and SMT components for solder masking fixtures.

ing surface-mounted parts in close proximity to related lead-type parts, the circuit connection is kept short. This allows for more efficient use of the top side of the PC board, leaving the opposite side of the substrate free for interfacing with other lead-type components. Details are shown in FIG. 5-10.

Mixing leaded components on the SMT assembly, of course, adds additional steps to the soldering process. Mixed technology will involve the use of additional fixtures and assembly equipment in an automated line. See FIG. 5-11.

Fig. 5-10. The low profile SMT devices will easily nest between the radius of large axial lead components to make maximum use of the substrate area and reduce circuit conductor length.

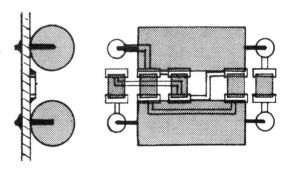

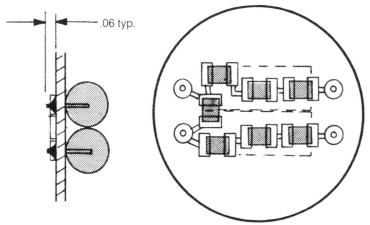

Fig. 5-11. Close coupling of chip components on the opposite side of the leaded device can reduce line inductance and increase component density.

CONNECTORS AND INTERFACE FOR SMT ASSEMBLY

Connector manufacturers have introduced several choices for surface-mounted applications. In addition to the more prominent names in connectors, several specialty products from domestic and offshore suppliers are providing unique interconnection systems. Selecting a connector that will mate with existing connector families is preferred, since this obviously provides flexibility and, in many cases, multiple sources (FIG. 5-12).

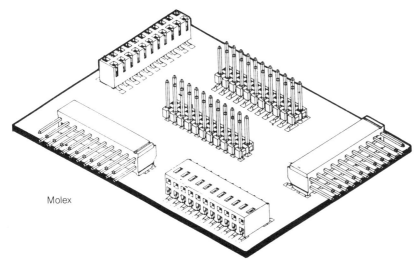

Molex

Fig. 5-12. Surface-mounted connector families with established mating cable systems will economically meet most interfacing requirements.

Connector manufacturers furnish retention or strain relief mounting tabs, bosses, etc., for the user who requires additional mechanical support beyond the solder connection. Techniques for mechanically retaining the connector to the substrate surface will vary from one manufacturer to another.

Heat Seal

Another method of interfacing assemblies used in consumer products is the heat/pressure seal flat cable (HSC). The heat/pressure seal flexible cable is a polyester film base material with parallel rows of graphite or graphite-silver conductors covered by an insulating layer.

The ends of this flat cable are free of insulation. When heat and pressure are applied, the electrical connection is made to the substrate.

Heat seal is a popular technique for mating the PC board to the glass surface of the liquid crystal display (LCD) components. Refer to HSC flexible cable manufacturers for specifications on compatible mating contact materials and environmental limitations. Details are illustrated in FIG. 5-13.

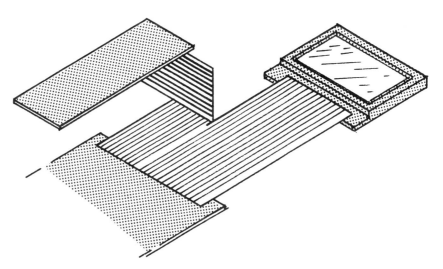

Fig. 5-13. Flexible cable systems can be terminated to several substrate types without the need of a connector.

Compression

Compression connectors are designed to join two or more parallel substrates that have matched contact areas on each mating surface as shown in FIG. 5-14. The substrates must be mechanically retained because this is usually how the connector material is captured within the assembly. If a more reliable method of retention is desired, a small amount of epoxy adhesive placed at the ends of the compression material will permanently attach the connector to one substrate surface or the other.

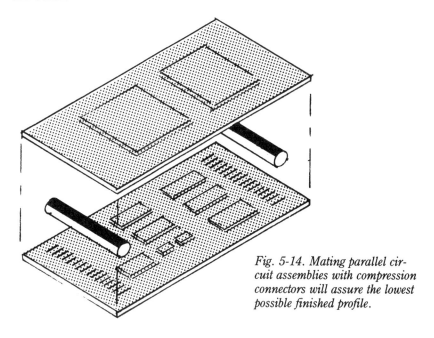

Fig. 5-14. Mating parallel circuit assemblies with compression connectors will assure the lowest possible finished profile.

6
Test Automation for SMT Assembly

PLANNING FOR AUTOMATIC TESTING

Planning for automatic testing of assembled PC boards begins in the initial stages of design. The type of testing is determined by a number of factors, including volume, product life, and equipment to be used.

Automated test equipment developed for SMT applications must accommodate higher component density and on occasion simultaneous testing of both sides of the assembly. Test equipment manufacturers have developed the *dual bed of nails* fixtures required to make contact with both sides of the circuit board simultaneously. These fixtures are three to five times more expensive than the conventional one-sided test bed and should be avoided. Examples are shown in FIG. 6-1.

TEST POINT CONTACT GUIDELINES

The test point should be clear of the component body and contact. Any side loading of the fragile test probe might damage the spring action required for proper contact pressure. When adding all tolerance limits of the PC board and test fixture elements, one may conclude that a contact area of .025-inch diameter would be adequate. Larger probe areas, .035/.040-inch diameter or square, for test points would be more reliable. The illustration in FIG. 6-2, represents a typical circuit with a test pad for each *net* or *node*. Note that all test points are clear of the component body, thereby reducing the danger of interference.

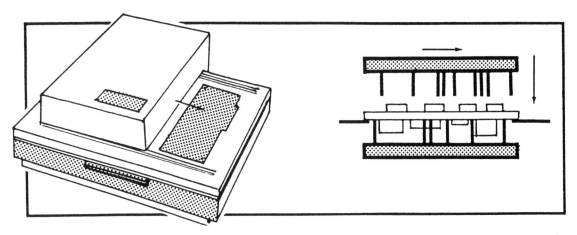

Fig. 6-1. In-Circuit Test of surface-mounted assemblies will require a single- or double-sided fixture with spring loaded probe contacts.

ASSEMBLY TEST METHODS

Many test levels are possible, depending on the environment in which the product operates. Three test methods popular for commercial and consumer applications are:

- In-Circuit Test (ICT): making test probe connection to each net or common connection of two or more components on the board;
- Functional Cluster: partitioning the PC board in a modular fashion with test points outside the component area;
- Board Level: using very refined test programs, the entire board is analyzed through connector interface using built-in diagnostics.

Test probing each contact point of every device can be expensive and will require additional real estate. Typically the "net" or common node of several components is more practical. A functional test of a partitioned cluster will speed up the test cycle and isolate problem areas to a specific area or component.

Board level testing will require more sophisticated programming to isolate the trouble area or the component that does not meet specification. This method requires refined equipment and engineering effort, but the hardware complexity is greatly reduced.

FIXTURE PLANNING

Tooling and handling fixtures for test equipment are a significant investment. Standardizing the board size reduces cost for each process

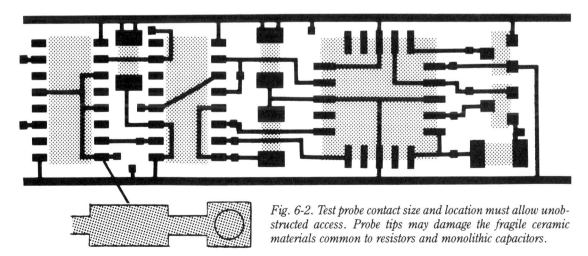

Fig. 6-2. Test probe contact size and location must allow unobstructed access. Probe tips may damage the fragile ceramic materials common to resistors and monolithic capacitors.

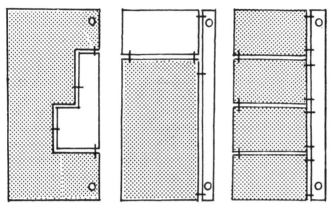

Fig. 6-3. Standard board or panel sizes with common tooling hole location can reduce basic fixture costs.

the assembly passes through. This does not mean the final outline of the board must be common, only that the rectangular blank should be of uniform size when possible. The designer and manufacturing engineer can work as a team to plan an efficient universal fixture that will keep costs under control. See FIG. 6-3.

TEST PROBE CONTACT IDENTIFICATION

The test engineer would be involved with the identification and numbering of test nodes on the schematic prior to the PC board layout. The X – Y position of each node should be furnished to the test department with the reference designator for each component. This data will be used in preparing test fixtures and the development of test software.

The test point contact area should be well clear of any component body and opposite the side of greater component density. To easily distinguish the test pad area from other via pads, consider using a different size or shape. For example, a round via pad could be used for general front-to-back interconnections and a square pad could be reserved for test probe points only, as shown in FIG. 6-4.

For those using CAD to design boards, a layer would be reserved to provide for test contact identification. Since a number is assigned to each node, the test point will become a part of the net and verified as a specific physical component.

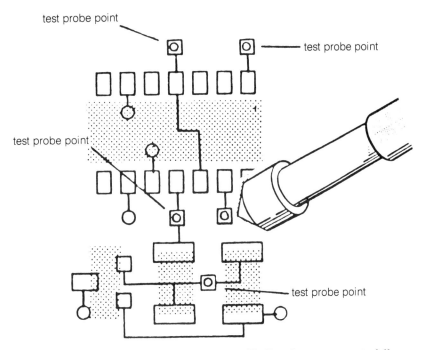

Fig. 6-4. One test contact or node per circuit net will allow the test systems to fully exercise each device on the assembly.

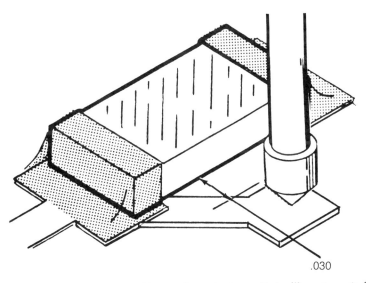

Fig. 6-5. The designer must provide a probe contact area that will compensate for the tolerance accumulation of the test fixture parts and a typical fabricated substrate.

PC BOARD TOLERANCE
AND FIXTURE PREPARATION

Tolerance capability of precision machinery and commercial PC board fabrication are quite different. It is common for PC board shops to profile the etched circuit with high-speed routing using a tracing template registered with locating pins.

The accuracy of outside dimensions could vary from the tooling hole locations up to .010 inch. On high-volume boards, it would be wise to consider die punching PC boards with hard tooling. The cost of this tooling is far greater than a routing template, but the accuracy and price per board in high volume can easily compensate for the expenditure. Machine tooling and fixtures are a more accurate process and afford closer tolerance capability.

Knowing what is possible and what to expect from test fixture services will help determine the location accuracy of the test probe contacts. The preferred test probe is designed for .100-inch spacing and is reliable, as well as economical. The miniature test probe contact can be mounted on .050-inch grid centers, but the fragile construction and high unit cost of the contact will restrict its use. Probe contact accuracy is determined by the quality of the equipment used to prepare the test fixture. The detail shown in FIG. 6-5 will compensate for the tolerance accumulation of the test fixture and a typical SMT circuit board.

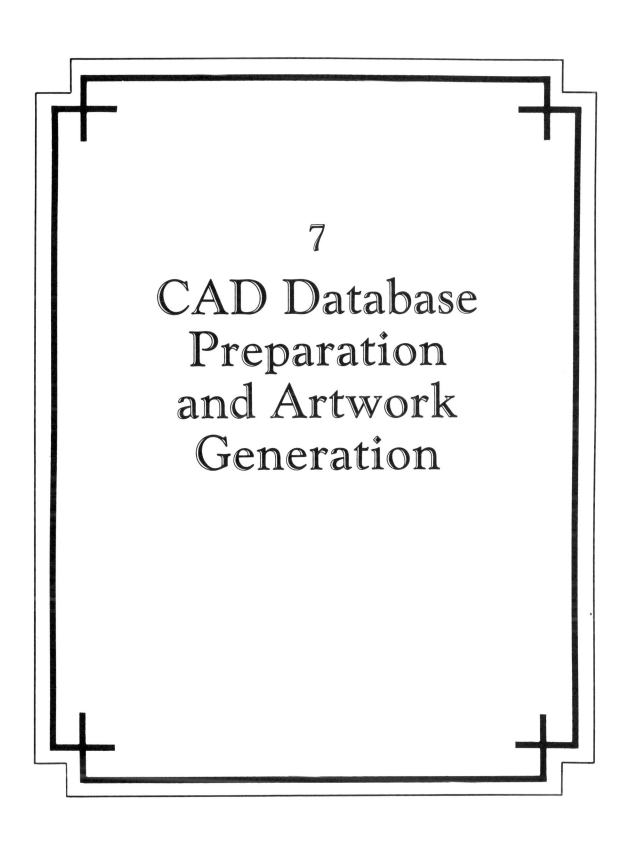

7

CAD Database Preparation and Artwork Generation

COMPONENT PARTS AND CONTACT PATTERNS WILL BE developed in the database of the CAD system in preparation for computer-aided design of the SMT circuit board. Grid positioning of the contact-land patterns will assist the CAD designer in efficient component placement and trace routing.

The recommendations shown on the following pages restate many of the contact patterns detailed in chapter 4, but dimensioning is redirected to the grid centers of the land pattern.

The JEDEC-registered general purpose IC packages have .050-inch spaces between lead centers. Component manufacturers have recommended standard limitations for the contact pattern (footprint), but are not familiar with the technology for creating artmasters for PC board fabrication. When preparing the footprint pattern recommendation, component suppliers are not aware of the importance of grid location in design.

Contact patterns must allow grid center placement when creating the footprint library in a CAD system and accommodate autorouting. FIGURE 7-1 describes the typical footprint pattern used for the JEDEC-registered plastic leaded chip carrier. The .025-inch contact width will allow for the routing of .008-inch wide conductor traces between leads when needed. If length is added to the contact pattern, it might be necessary to reposition the grid center in order to maintain the desired solder reflow characteristics.

FEEDTHROUGH AND VIA PADS

When placing plated through-hole pads on the design, do not allow the pad to make direct contact with the component contact area. Clearance between the footprint pattern and feedthrough pads must allow a

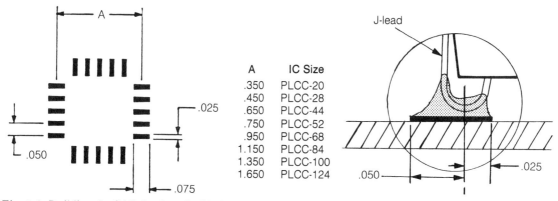

A	IC Size
.350	PLCC-20
.450	PLCC-28
.650	PLCC-44
.750	PLCC-52
.950	PLCC-68
1.150	PLCC-84
1.350	PLCC-100
1.650	PLCC-124

Fig. 7-1. Building the CAD database for SMT land pattern families is easier if the contact geometries are referenced to a common grid center.

barrier of solder mask coating that will stop the migration of liquid solder during the reflow process. The conductor trace connecting the contact area with the feedthrough pad should be approximately .010 inch wide. See FIG. 7-2.

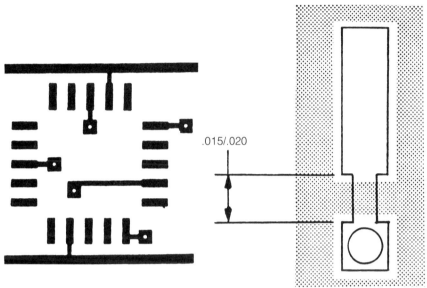

.015/.020

Fig. 7-2. The space between land pattern and via hole and pad features must allow for a solder mask barrier to restrict the solder to the contact area.

Feedthrough pads against or within the footprint pattern as shown in FIG. 7-3 will cause a migration of liquid solder from the contact area during reflow. Without sufficient solder on the contact of the component pad, parts will lift as the solder cools, causing a serious defect.

Via holes in the SMT contact area are not recommended. The added cost of the small hole size and plugging processes is not a desirable alternative to the separation of contact and feedthrough pads. If it is necessary to design feedthrough pads directly against contact areas or connect feedthrough pads with contact areas with heavy traces, consider the following options: solder mask over bare copper, solder mask over feedthrough pads, or plug the feedthrough hole with plating (plate closed).

To further provide for routing conductor traces while ensuring an acceptable air gap, use a .035- to .040-inch diameter or square pad for all .018/.020-inch diameter feedthrough holes. The square configuration furnishes more than enough metal in the diagonal corners to compensate for the reduced annular cross-section at the sides of the square. When

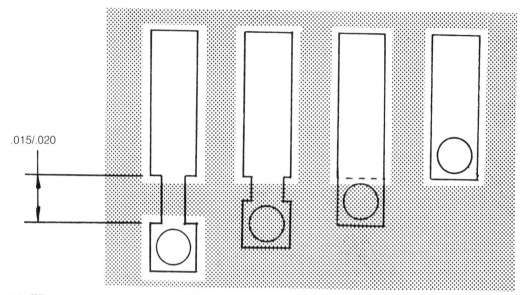

.015/.020

Fig. 7-3. When a via hole and pad is located too close to the contact area, it will be necessary to cover or tent over the pad area with solder mask.

identifying a via as a test probe contact area, use a .035- or .040-inch square pad. Provide a minimum space of .100 inch between probe contact locations. See FIG. 7-4.

Do not position via pads under small resistors or capacitors. Migration of solder during the reflow process will cause the liquid metal to draw away from the contact area. Details shown in FIG. 7-5 further illus-

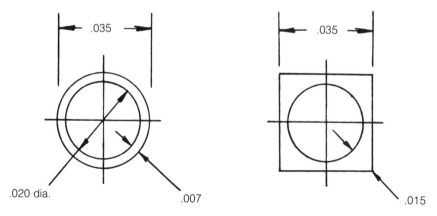

.035

.035

.020 dia.

.007

.015

Fig. 7-4. A .010-inch annular ring around a finished via hole is preferred, but due to space restriction, it may be less. By furnishing a square shape for via hole pad, the reduced annular ring is confined to a very small section of the overall area.

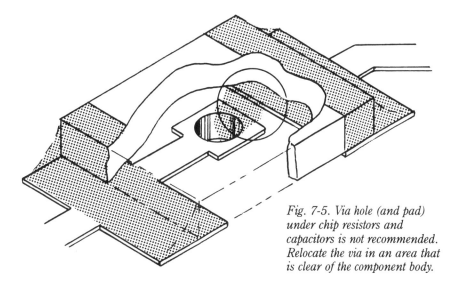

Fig. 7-5. Via hole (and pad) under chip resistors and capacitors is not recommended. Relocate the via in an area that is clear of the component body.

trate the danger of solder bridging under the body of the chip component. This can occur if a small amount of solder paste is lodged under the component or during post wave-solder operations.

Contact patterns too close to feedthrough pads will cause unwanted production problems. The solder paste, when heated to a liquid, flows into the feedthrough hole and away from the contact area. It is common to position via holes and pads under the bodies of ICs. Avoid hidden feedthrough pads near unrelated contact patterns. The only way to correct a bridge or short under the component is to desolder and remove the IC.

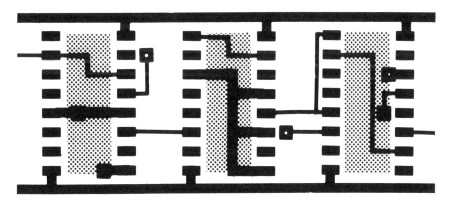

Fig. 7-6. Spacing between the via hole pad and device land patterns, if not covered by solder mask, may bridge during assembly processing.

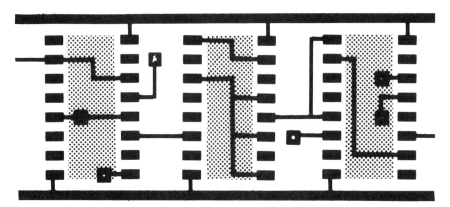

Fig. 7-7. By adhering to minimum spacing recommendations on via hole pads, the designer reduces the need for costly rework to remove solder bridging.

Adding feedthrough holes directly onto the contact areas is also allowable if the hole itself can be plugged to stop solder paste migration. Solder paste migration and bridging are to be avoided. Rework and touchup are unwanted detours in the high-volume manufacturing operation. Using these guidelines will minimize problems during the assembly process.

Illustrated in FIG. 7-6 are potential problems with feedthrough pads on SOICs, while FIG. 7-7 compares good design practice, with clearance provided for solder mask, to a poorer design practice that will require tenting or covering the via holes with solder mask.

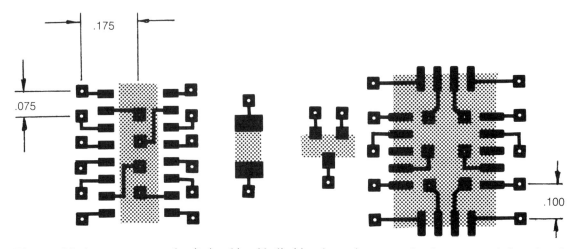

Fig. 7-8. Maximum component density is achieved by limiting the surface area to land patterns and via pads only. All signal interconnection is made on internal layers of the substrate.

As component density increases further, it may be necessary to limit the outside layer surfaces to component-mounting contact patterns and feedthrough pads only. See FIG. 7-8.

The reduced size of the plated feedthrough hole diameter easily allows three conductor traces on internal layers without resorting to *fine line* (.006-inch-wide) traces.

SURFACE-MOUNTED DEVICES ON
SIDE TWO OR WAVE-SOLDER SIDE

The chip component patterns shown are recommended for reflow-solder processes, but also work well on dual wave-solder machinery. Because wave soldering is one of the most popular processes in the world, many companies choose to use leaded through-hole components for all ICs and larger passive devices. By mounting the majority of resistors, capacitors, and SOT components with epoxy adhesive on the wave-solder side, board size can be reduced. Mounting chip components on the wave-solder side, however, will restrict conductor routing paths. See FIG. 7-9.

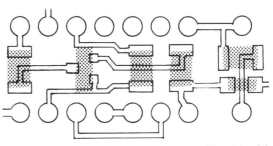

Fig. 7-9. Nesting surface-mounted devices on the wave-solder side of the substrate will increase component density, but restricts circuit routing paths.

The footprint patterns illustrated in FIG. 7-10 allow for placement accuracy of the assembly equipment, as well as for possible component shift during adhesive cure. When using the narrower contact pattern, be sure to allow adequate clearance between the component bodies, conductor traces, and vias. Keep in mind that the component body will overlap both sides of the footprint pattern when using this design.

Many footprint design possibilities exist that further reduce secondary rework procedures. Component footprint patterns have been designed to reduce or eliminate rework and touchup of boards after assembly procedures. FIGURE 7-11 compares alternative contact geometry used for wave-solder application.

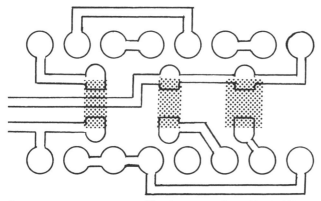

Fig. 7-10. A narrow contact geometry will reduce excessive solder buildup on a component during wave-solder connections.

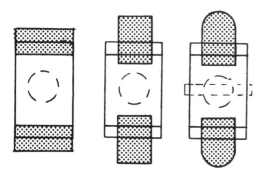

Fig. 7-11. The three examples of chip component land pattern options are for epoxy attachment and wave solder. Excessive solder buildup on the component end termination will cause stress-induced cracking.

BUILDING CAD DATABASE FOR CHIP RESISTORS AND CAPACITORS

The footprint patterns dimensioned in FIG. 7-12 will accommodate .025-inch grid positioning, generally compatible with CAD systems. The details shown are for the full-width general purpose style of pad geometry that is allowable for reflow- or wave-solder process.

For those companies that require a visible side castilation of solder on capacitors, .010 inch can be added to the ''B,'' or width, dimension.

OPTIONAL CAD DATABASE FOR WAVE SOLDER

The contact geometry, detailed in FIG. 7-13, is specifically designed for epoxy attached components that will be processed in a wave-solder system.

Caution: Compensate for the component body and contact area when using the narrow footprint pattern. When spacing the device, allow

102

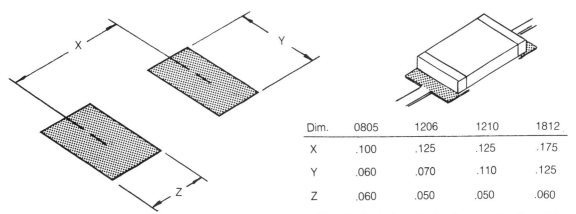

Dim.	0805	1206	1210	1812
X	.100	.125	.125	.175
Y	.060	.070	.110	.125
Z	.060	.050	.050	.060

Fig. 7-12. Center line location for SMT land pattern features will assist the designer in database construction of the CAD library.

for an additional .040-inch clearance to an adjacent component or feed-through hole. Wave solder of larger chip components—1812 and 2225—is not advisable due to the high occurrence of cracking caused by thermal mismatch of the substrate and ceramic dielectric.

TANTALUM CAPACITOR AND CONTACT GEOMETRY

Molded devices such as tantalum capacitors and inductors are currently available from multiple sources. Careful selection of a reliable supplier for tantalum capacitors is the safest route to follow. Compare the

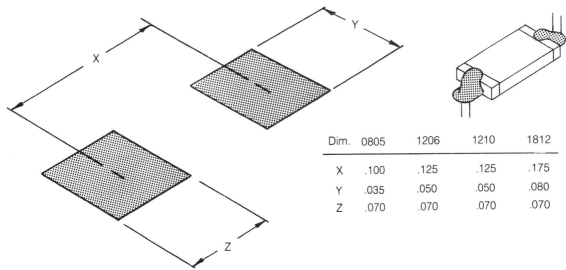

Dim.	0805	1206	1210	1812
X	.100	.125	.125	.175
Y	.035	.050	.050	.080
Z	.070	.070	.070	.070

Fig. 7-13. The optional narrow land pattern for chip resistors and capacitors should be used only for wave solder of epoxy attached components.

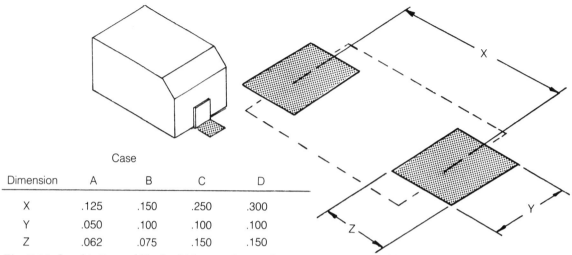

Dimension	Case A	B	C	D
X	.125	.150	.250	.300
Y	.050	.100	.100	.100
Z	.062	.075	.150	.150

Fig. 7-14. Land pattern width should be very close to the component contact width. Excessive area of the patterns will cause component shift during the reflow-solder processing.

choice with the standard tantalum configuration and value/voltage range in chapter 3.

The contact geometry illustrated in FIG. 7-14 will work well with the EIA standard tantalum capacitor in A, B, C, and D case sizes.

SOT-23 Contact Geometry

FIGURE 7-15 is a pattern for the MMD/MMT or *micromin* transistor device, but can be used as a universal pattern which will also accommodate the SOT-23. The important features to be noted are the grid positions of the contacts.

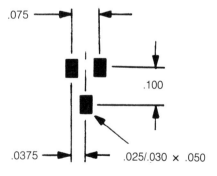

Fig. 7-15. To simplify CAD database construction of the SOT-23 land patterns, one rectangular shape for all three contacts can be substituted for the pattern recommended in chapter 4.

SOIC Contact Geometry

The JEDEC-registered small outline, or SOIC contact pattern is easily adaptable to CAD layout. The two parallel rows of contacts, as stated in chapter 4, have the same pin assignment as the dual-in-line, through-hole IC it replaces. For most logic devices, the .200-inch spacing between contact-row centers will be consistent. Some 16-pin devices require a wider lead frame to provide for the die size or heat dissipation. A PC board may have a mix of the SO-16 and SO-16L (wide) devices. SOICs greater than 16 leads will always be in the wide lead frame, but they will be limited to a total of 28 leads. Most active components, beyond 28 leads, will be supplied in the PLCC or other quad configurations. (See FIG. 7-16.)

SO-8, 14, 16   SO-16L, 20 - 28

Fig. 7-16. The land patterns furnished for the JEDEC registered SOIC family of devices conform to the IPC-SM-782 standards.

Active components in the SO package can be obtained from multiple sources and will be dimensionally and pin-for-pin compatible with a device manufactured in another part of the world.

The first suppliers of the SOIC chose the .050-inch lead spacing to be compatible with the accepted grid system. The quad arrangement on plastic chip carriers can also be located on a grid pattern.

While Asian manufacturers use the metric system, they do maintain the .050-inch spacing on many components. But among products available from Asia with .050-inch lead spacing, there are significant mechanical differences. The distance between rows of contacts on the EIAJ-SOP IC is not the same as on the JEDEC-SOIC. If it is necessary to use IC devices from multiple sources, it may not be possible to mount both offshore and domestic sizes using the footprint pattern recommended by the component manufacturer.

FIGURE 7-17 shows manufacturer's differences in footprint size and row spacing. Even though each device is pin-for-pin compatible, the designer must choose one or the other, or develop a universal pattern to accommodate them.

The SO/SOP universal pattern is a configuration to consider. Because of the excess length of the contact area, the IC may tend to shift to one side during the solder reflow process. If this slight shift is not acceptable, it would be necessary to add adhesive to the component when the IC is placed on the board surface. Many Asian component suppliers do comply with the JEDEC standard and offer direct compatibility with domestic manufacturers.

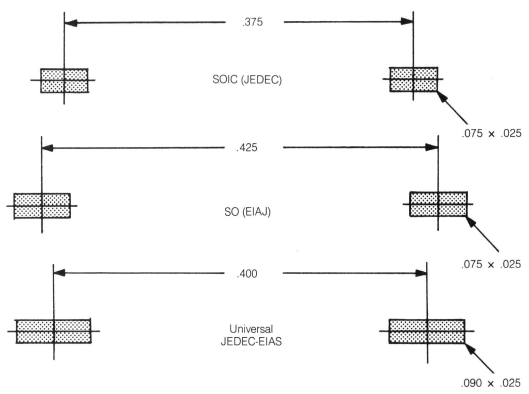

Fig. 7-17. Because many Asian sources for small outline IC devices do not conform to JEDEC standards, it is often necessary to adopt a universal land pattern.

CONDUCTOR TRACE ROUTING

When preparing for photoplotting of closely spaced conductor traces on an SMT board, it is essential to reduce trace width. Routing wide signal lines between the .025-inch-wide contact patterns used on most SMT ICs requires reducing the conductor width to .008 inch overall or, as traces pass between contact patterns, providing an air gap.

When necking down, a smooth transition from one width to the other can be achieved if each signal trace starts and stops at the same grid point. This stopping point might not be obvious when looking at a pin plot, but the careless overlap of line connections can cause potential problems when the photoplot is generated.

Take care to overlap the start- and stop-position of the line aperture when photo-plotting. Line apertures of photo-plotters and laser plotters are generally a circular opening, so the end of any line run will have a full radius end. Because CAD systems can snap to a grid point, it is customary to use a 45-degree angle when traces must diverge from a continuous line and continue to travel in the same direction. Offset-stepping several conductor traces on close spacing will require attention in order to maintain a proper air gap. The start- and stop-points of these aperture runs must be carefully executed, reducing the chance of overlap and shorting. Common pitfalls and limitations of aperture photo-plotting are eliminated when laser plotting is used.

Some designers choose to reduce the contact patterns for SOICs and PLCCs to .020-inch width (FIG. 7-18). This allows the use of .010-inch conductor trace width throughout the board and necking down is reduced to a minimum. While CAD-routing and photo-plotting time is reduced, the cost of the PC board can increase due to the finer line etch control necessary. The .020-inch-wide contact can cause additional difficulty in placing larger PLCC ICs. Robotic equipment that is not

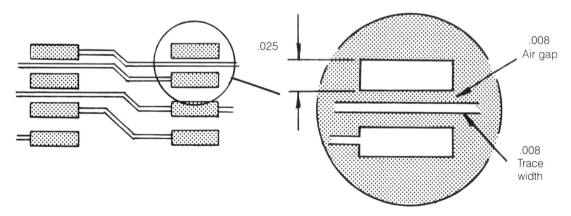

Fig. 7-18. Conductor routing between .050-inch spaced land patterns is common. The land pattern geometry to conductor trace spacing should allow for a minimum air gap of .008 inch.

equipped with optical recognition will have difficulty aligning with this narrow target. Placement and rotational accuracy must be 25 percent greater due to the reduced contact area.

ROUTING SMT

High-density autorouting for surface-mounted PC boards is not as efficient with most CAD systems as autorouting of PTH boards. The systems are primarily hole-intelligent through all layers of the multilayer board and cannot route to the SMT contact patterns efficiently.

For maximum efficiency of a CAD system, the designer must create a footprint library that will include the component contact pattern and a uniform, pre-assigned via pad connected by a narrow trace (.008/.010 inch wide). The via pad will appear on all layers, but the contact pattern and the trace appear only on the outside layer. FIGURE 7-19 illustrates the advantage of the pads-only configuration. The inside layers will provide unrestricted routing paths between vias and under contact areas. The router can now ignore the footprint of the component and address only pre-assigned via pads. Of course, via pads added by the router program must be kept out of the footprint zone.

To further increase density, especially when components are mounted closely on both sides of the board surface, the designer could have to revert to blind or buried vias as shown in FIG. 7-20. This procedure should be avoided because of the difficulty and added cost to board fabrication.

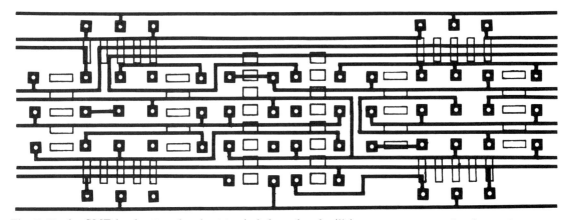

Fig. 7-19. An SMT land pattern breakout to via hole and pad will increase component density on the substrate surface. All conductor trace interconnection would be shifted to internal layers for efficient CAD auto-routing.

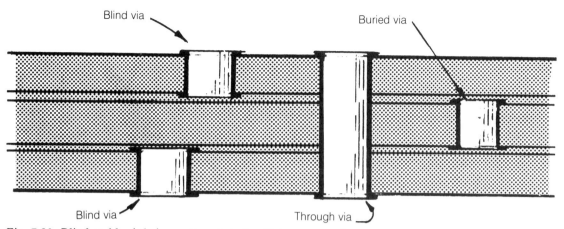

Fig. 7-20. Blind and buried vias on the substrate will increase conductor trace routing density, but fabrication is difficult and costly.

Furnishing manufacturing personnel with X – Y coordinates of components will accelerate the implementation of CAD/CAM necessary for automation.

As the footprint library is being prepared for each device, add the center point of the component on one of the pad stack layers. This center-point position will be referenced from the zero datum of the PC board. The datum will easily reference the offset of any individual assembly system.

When orientation of the component is supplied, programming of the equipment will take only a fraction of the time required to make X – Y calculations from film masters or board models. See details in FIG. 7-21.

HAND-TAPED ARTMASTER PREPARATION

Generating artwork for the two-sided SMT PC board is similar to artwork supplied for the conventional leaded PTH board. The major difference is separation of the SMT footprint patterns. When choosing pre-printed footprint patterns for surface-mounted designs, make sure the pad geometry is process-proven or meets the recommendation of the component manufacturer or IPC-SM-782. The scale for preparing SMT artwork is 4-to-1 or greater to ensure accuracy and maximize density.

One sheet or layer of mylar is dedicated to the footprint patterns of SMT components. The footprint pattern layer, when photo-reduced to 1:1 size, will be the master for preparing the solder paste screen. This film is not required if the wave-solder process is used.

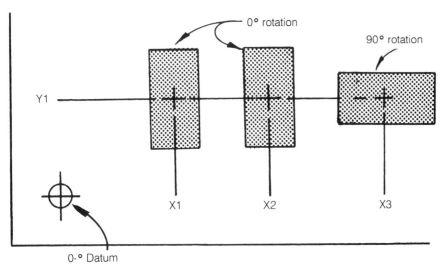

Fig. 7-21. Constructing the component land pattern database from the center of SMT device is preferred. This will provide manufacturing with the location of orientation for each part from a reference datum or fiducial target.

ARTWORK LAYERS: 2-SIDED PCB, WITH SMT ON ONE SIDE

- Pad master (all common vias, boarder, etc., sheet 1)
- Footprint master (all SMT component patterns, sheet 2)
- Side-one conductor (sheet 3)
- Side-two conductor (sheet 4)
- Side-one legend (screen master, sheet 5)

When ordering working film, instruct the photography service to furnish the following composite film set:

- Composite sheet 1, 2, 3 (side one)
- Composite sheet 1, 4 (side two)
- Composite sheet 1, 2 expanded .010 inch (side one, solder mask)
- Sheet 1 only expanded .020 inch (side two, solder mask)
- Sheet 2 only (solder paste screen master)
- Sheet 5 only (legend screen master)

All the film except sheet 2 (the solder paste master) is sent with a fabrication drawing to the board vendor. The solder paste master is only used to prepare the screen or stencil used in applying solder paste for reflow process.

All areas of the board except the contact areas and feedthrough pads will be coated with solder mask. The solder mask is necessary to contain solder paste on the contact pads. Board fabrication and substrate options will be expanded on further in chapter 8.

8

Specifying Substrate Materials and Fabrication Options

G LASS-REINFORCED RESIN LAMINATE (FR-4) HAS, FOR many years, been the primary substrate material for etched copper circuit boards. FR-4 has proven ideal for two-sided and multilayer PC boards that use plated through-holes to interface one layer with another. The fire-retardant material has become the industry standard for communications, instrumentation, computers and virtually all electronic products requiring high quality performance.

There are alternatives to FR-4 in electronic applications. Low-cost consumer products use one of several paper-based laminates on the market. Etched circuits with simple functions might require copper conductor traces on one surface. Holes for component leads can be punched, rather than drilled through the substrate as shown in FIG. 8-1.

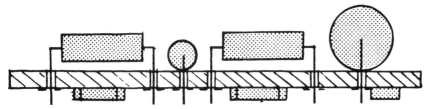

Fig. 8-1. SMT components are attached to the etched copper side of the substrate with epoxy for wave-solder processing.

COMMON SUBSTRATE MATERIALS

- FR-2 Low-cost, flame-retardant paper, the workhorse of consumer electronics when electrical and mechanical properties are not overly demanding.
- FR-6 Glass-mat, reinforced polyester. Low-cost favorite of the automotive industry.
- CEM-1 Paper-glass composite. Best laminate property for price ratio. Can be used to replace other laminate grades to improve yields, cut laminate costs, or obtain better process. (Example: replaces single-sided FR-4.) Extensive use in radios, smoke detectors, and other consumer applications.
- CRM-5 Composite material using a polyester resin system to bind together a sandwich of random glass mat core and woven glass fabric surfaces.
- CEM-3 All glass composite. A recent entry into the laminate spectrum. Cheaper than FR-4 with almost equivalent properties. Entering automotive and appliance applications.
- FR-4 Fire-retardant glass laminate, more expensive than the above materials, but very good. Widely used in computer and telecommunications applications.

In addition to the commercial materials listed, the designer/engineer can handle more demanding environments with one of the "high-tech" materials now available.

Teflon and ceramic substrates are used in many RF applications requiring a stable dielectric characteristic. The teflon materials are furnished copper-clad, as are the more conventional laminates. Teflon is processed in a method similar to FR-4.

Ceramic substrates are fabricated using very different techniques. The substrate area is limited and conventional drilling or routing of this material is impossible. The ceramic substrate can be punched and profiled in its green or unfired state. After firing of the organic material, holes and additional alterations to its shape will require laser technology. The circuit traces are added to the surfaces of the ceramic layers and conductive fillers are drawn into the laser drilled via holes to connect one layer to the other.

Ceramic substrates are used primarily for hybrid circuits. In many cases, IC chips are attached directly to the surface of the ceramic and terminated using wire bonding techniques similar to those used in IC fabrication. The advantage in using a hybrid circuit is the choice the engineer has in mixing several integrated circuits, transistors, and miniature surface-mounted devices onto the substrate. The result is a customized electronic function in a small package. The finished module may have leads attached directly for mounting into a larger PC board or, as in *hi-rel* applications, sealed in a metal housing with glass insulated leads. See FIG. 8-2.

Other alternatives to the more common laminates are polyimide-glass and polyimide-quartz materials. Polyimide products, more stable than FR-4, can be processed with most of the techniques used in standard board fabrication. Because the polyimide has a more stable thermal coefficient of expansion than FR-4, it is popular for chip-on-board applications similar to those used in hybrid circuits. Polyimides can be laminated to metal core layers to provide an even more thermally stable substrate surface. Details are shown in TABLE 8-1. This application will be discussed in detail later in this chapter.

Note: Contact the board fabricator to assist in writing specifications when board material calls for special requirements. This will eliminate misunderstandings, which could delay processing of the product.

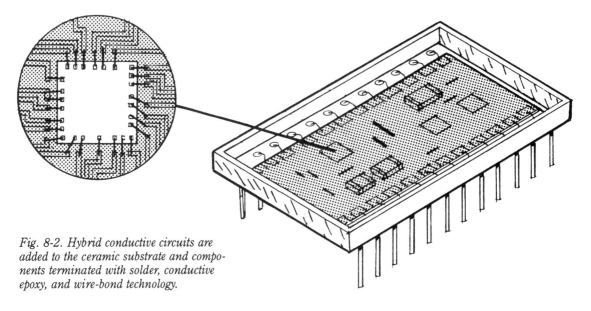

Fig. 8-2. Hybrid conductive circuits are added to the ceramic substrate and components terminated with solder, conductive epoxy, and wire-bond technology.

Table 8-1. Dielectric Constants and Thermal Coefficients of Expansion for Polyimide and Related Materials.

Material Type	Dielectric Constant	Thermal Coefficient of Expansion
Epoxy Glass	5.0 – 6.0	15.8
Polyimide	4.3 – 5.0	14.2
Polyimide/CEC*	4.3 – 5.0	6.4
Ceramic (alumina)	8.5 – 9.5	6.4
Epoxy Kevlar	3.8 – 4.5	6.5
Epoxy Quartz	3.5 – 4.0	6.0

*Copper invar copper metal core.

FABRICATION AND MATERIAL PLANNING

The bulk of commercial electronics will continue to use epoxy-glass laminate, while paper-epoxy will be confined to consumer products.

Substrate materials are furnished to the board fabrication in 36-×-48-inch sheets. The fabricator then cuts the material into smaller panels for processing, i.e., 18×24 inches, or 12×18 inches. From these panels, the PC board is processed into one, two, or more PC units per panel. The fabricator plans material usage carefully for minimum waste.

If the designer of the board prepares film in multiple image format, two areas are to be considered: the maximum size board the assembly equipment can handle, and minimizing waste of the board material. See FIG. 8-3.

Routing to profile the board is one of the last steps in the fabrication process, but planning must be done in advance to ensure the best results.

Methods of profiling the individual board shape include: rotational cutters, high pressure hydro-routing, laser profile, bevel scoring, and shearing after assembly. For the latter, adequate clearance to component bodies is necessary for the straight cuts.

The slot separating the breakaway tab from the individual boards on a panel should be a minimum of .100 inch wide. Generally a 0.090-inch diameter router tool is used to create the 0.100-inch wide slot in two passes to ensure greater tooling hole-to-board edge accuracy. The tabs

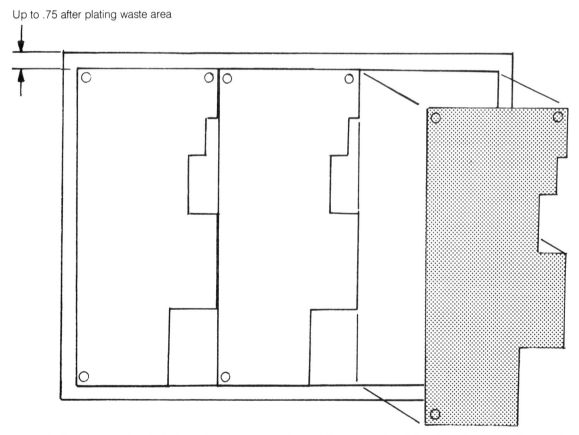

Up to .75 after plating waste area

Fig. 8-3. Several irregular shaped substrate images can be supplied on one panel for efficient automated assembly machine handling.

retaining the board to the panel are usually .100 inch wide and are spaced to adequately support the panel as a unit through the assembly process. There are several variations used for the connecting tabs. FIG-URE 8-4 illustrates a few.

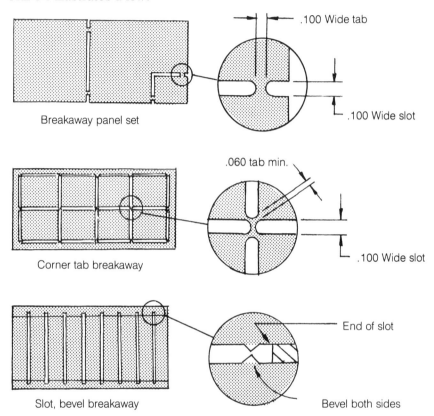

Breakaway panel set

.100 Wide tab

.100 Wide slot

.060 tab min.

.100 Wide slot

Corner tab breakaway

End of slot

Slot, bevel breakaway   Bevel both sides

Fig. 8-4. Several techniques for retaining substrate units in a panel form are available. Each of these examples represents an optional routed slot pattern.

Die cutting or punching methods are normally reserved for higher volume products and have proven to be efficient means of fabrication, while glass laminates are less than ideal for this method. Due to the abrasive nature of the glass fiber material, paper-base laminates are excellent. Punch-press profiling of the board (FIG. 8-5) can be configured similar to a routed panel, but processed with greater speed and economy.

Panelizing small boards is a must for efficient assembly processing. FIGURE 8-6 illustrates an example of an irregularly shaped board that is awkward to handle as an individual unit.

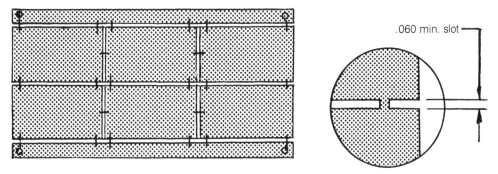

Fig. 8-5. Die punched slot and tab methods require a hard tooling investment, but can reduce panel fabrication costs significantly over the life of a product.

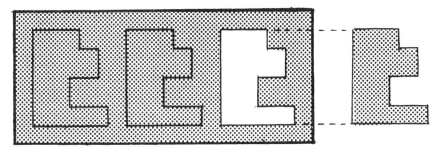

Fig. 8-6. Small and irregular shaped substrates can be retained in a panel for easy, post-assembly excising.

Small profiles like this can be blanked from a panel in one hit, much like a cookie cutter. Then each board is pressed back into the panel. This method provides efficient handling through all assembly operations. The final operation after assembly requires simply pushing the finished module from the panel. This concept is referred to as the push-back or punch-and-retain panel. Adequate clearance between each module must be provided for the male/female die set contact with the panel surface.

PROVIDING FOR ASSEMBLY AUTOMATION

Before preparing the fabrication detail, the designer must know the assembly equipment that is to be used. Basic information would include maximum panel size, tooling-hole requirements and placement area for the surface-mounted devices.

There are many equipment choices on the market. When the right combination is selected, all assembly stations will be compatible.

Before going to the expense of panelizing your product, a design

Equipment	Placement Area	Maximum Board
Dynapert MPS 500	16 × 18	18 × 20
Quad QS-34	18 × 18	20 × 20
Fuji FP 60/90	9 × 13	10 × 14
Fuji CP 2	13 × 17	14 × 18
Excelon CP30/40	14 × 16	16 × 18
Amistar SM2001	16 × 20	16 × 16
Universal 4621	14 × 16	16 × 18

Table 8-2. Maximum Panel Sizes and Placement Areas for Surface-Mounted Components of Different Equipment Manufacturers.

review of an assembly by the manufacturing engineer would allow for improvements. This could save time and money over the life of the product.

The maximum panel size and placement area for pick-and-place of surface-mounted components varies from one equipment manufacturer to another. Listed in TABLE 8-2 are a few examples.

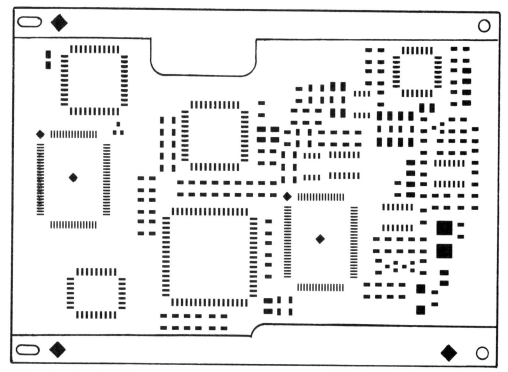

Fig. 8-7. Panel construction of one or more units-per-board should allow for machine automation of all assembly processes.

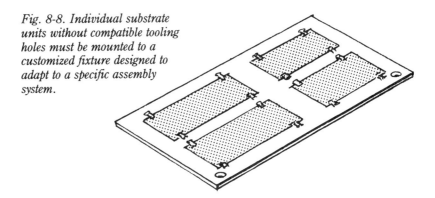

Fig. 8-8. Individual substrate units without compatible tooling holes must be mounted to a customized fixture designed to adapt to a specific assembly system.

When designing the board, furnish at least two holes for tooling location pins on each unit. In the case of panelization, two additional tooling holes are required, generally on a breakaway tab. The breakaway tab and holes will be discarded after assembly processes are completed. The tooling holes in each board will be used in text fixtures on the finished assembly. The tooling hole pattern shown in FIG. 8-7 allows for placement of SMDs on both sides of the board.

For long production runs, tooling holes on the board should be compatible with all machines in the assembly process from screen printing, pick-and-place and solder reflow to cleaning stations. However, since one size circuit board is not practical for all products, universal carrier pallets are developed to transfer one particular assembly through each phase of the process. This is usually the alternative when smaller boards are not panelized. Two or more substrates will be attached to a pallet and pass from one machine to another as a set, as shown in FIG. 8-8. The advantage to this concept is that there is no downtime necessary for adjusting rails, tracks, or belts.

Other factors in preparing for automation include component orientation, clearance, or spacing between devices. These factors were discussed in chapter 1.

Clearance of board-edge to the component body is another point to be addressed by the designer. If the assembly is to pass over a wave-solder system, an adequate distance from components must be provided in order to eliminate interference with holding rails or pallet fixtures. Breakaway edge strips could be the best way to provide edge clearance on boards having high component density. See FIG. 8-9.

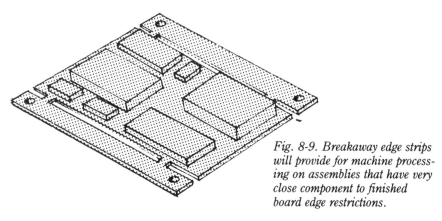

Fig. 8-9. Breakaway edge strips will provide for machine processing on assemblies that have very close component to finished board edge restrictions.

PLATED THROUGH-HOLES

Leaded through-hole parts continue to be a viable part of the electronics industry. The guidelines for plated holes of leaded parts are well established.

Component lead diameters will vary over a wide range. To reduce drill size variables and fabrication cost, PC board manufacturers advise standard hole sizes. For the majority of through-hole components, the lead diameter falls in the range of .018 to .028 inch. A common hole size can be selected if allowances are made for the tolerance of lead-forming equipment, automatic insertion accuracy and wave-solder characteristics. The industry has generally accepted the .037/.042-inch diameter hole range as a standard for auto-insertion application. Smaller holes can usually be adapted for hand assembly.

The .037/.042-inch size hole provides .010- to .012-inch clearance for the larger diameter leads supplied on resistors and capacitors, while at the same time compensating for the more difficult alignment of DIP ICs.

MULTILAYER AND FINE LINE CONSTRUCTION

The industry is pressed to push state-of-the-art fabrication technology as component density increases and circuit complexity evolves.

When the trace width and air gap are less than .012/.010 inch or plated through-hole diameters are less than .018/.020 inch in diameter, the increased difficulty of manufacturing is reflected in board cost.

Before reverting to the more difficult and costly high-tech method of board manufacturing, the designer should make an effort to keep fabrication less complex. The board is usually the single most expensive component of the assembly; therefore, savings at this level are reflected over the entire project and product life.

Most manufacturers of fixed-space leaded components have complied with the .100-inch grid arrangement. The .100-inch grid pattern has been used for many years as a standardized form with PC board designers. It is a common practice to use a conductor trace between .015 to .012 inch for most signal-carrying applications. Routing a conductor trace between .100-inch spaced holes will further restrict the outside diameter of the contact pad. The conductor path must be spaced away from non-related contact pads and conductors, maintaining the air gap or clearance. This air gap will allow for a clean etch during fabrication of the PC board and reduce the chance of solder bridging during soldering processes. See FIG. 8-10.

The tolerance of a drilled and plated hole is typically + .003 and − .001-inch diameter. Keeping the above mentioned tolerances and limitations in mind, .060 − .062-inch outside diameter contact pad size is recommended for leaded components requiring the .100-inch grid pattern.

Component density is further increased when applying multilayer technology. Typically the contact pad appears on the outside layer, and on an inside layer only when a connection must be made. Conductor density can be increased because the plated through-holes do not have

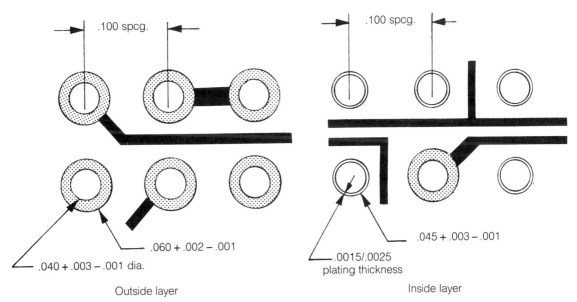

Fig. 8-10. Narrow conductor trace width on internal layers of the multilayer PC board will improve routing density.

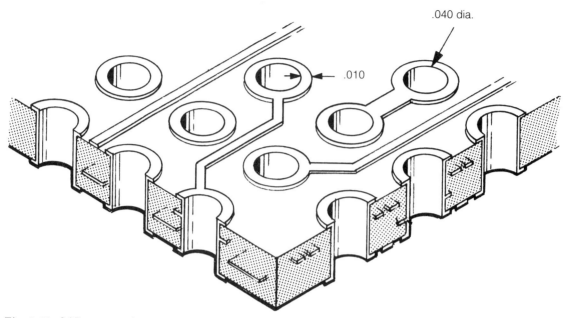

Fig. 8-11. CAD auto-routing of conductor traces is more efficient when via holes are located on a fixed grid pattern.

the annular ring on the inside layers. Two .010-inch wide conductor traces will provide a .010-inch air gap on these inside or laminated layers. The importance of accuracy in pad location and conductor trace spacing cannot be over-stressed. Many companies rely heavily on CAD systems designed to alleviate these complex applications. The .100-inch grid pattern is also the primary locating element in the auto-routing feature of more advanced systems. FIGURE 8-11 provides an example.

SURFACE MOUNT AND VIA HOLES

The surface-mounted component does not require a hole for each contact in the substrate. On a smaller, less complex surface-mounted circuit, it is possible to design the PC board without using any holes.

As SMT circuits become more complex, with increased component density, the need to add feedthrough (via) pads in the substrate to maximize available space also increases. Because feedthrough holes do not have to clear a component lead, a smaller drill size is possible.

In hybrid technology using ceramic materials, via holes are smaller than .010-inch diameter. However, in a glass laminate fabrication, small drill size will add excessive cost to the fabrication of the PC board. To maximize drill speed and to keep the drill breakage rate low, most board

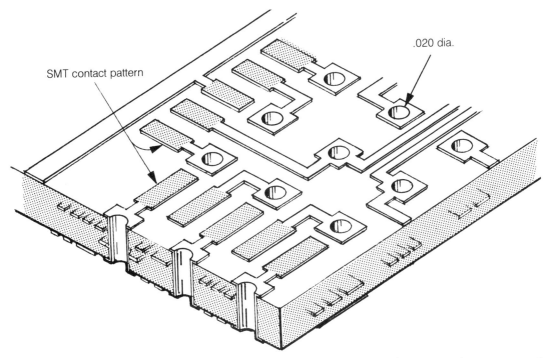

.020 dia.

SMT contact pattern

Fig. 8-12. Component land pattern breakout to the grid spaced via hole and pad improves conductor trace routing paths and provides for solder mask separation.

shops would prefer a minimum finished hole size of .018 to .020-inch diameter. This size hole allows for a reduction in pad size and increased conductor trace density. Standardizing on a .020-inch finished hole diameter will allow for a .040-inch outside diameter pad while maintaining the desired .010-inch annular ring. See FIG. 8-12.

To further provide for routing conductor traces while at the same time ensuring an acceptable air gap, the designer may choose to use a square pad of .035/.040 inch for feedthrough holes. This square configuration will furnish more than enough metal in the diagonal corners of the pad to compensate for the reduced annular cross-section at the sides of the square. The .035 – .040-inch square feedthrough pad can be spaced at .050 inch when necessary or on the more traditional .100-inch grid. When the square pad is spaced at .100 inch, it is possible to route two or three conductor traces between pads. Originally, this density was possible only on internal layers of multilayer boards with leaded through-hole technology. Examples are shown in FIG. 8-13. Using multilayer for surface-mounted applications will dramatically increase density possibilities.

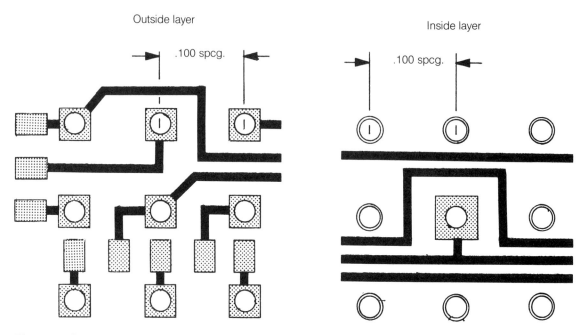

Fig. 8-13. Conductor traces of less than .010 inch in width on internal layers can take advantage of the space occupied by the pad around each via hole on the outside layers.

The reduced size of the plated feedthrough hole diameter will easily allow three conductor traces on internal layers without resorting to fine line (.006 inch wide) traces. As component density increases further, it may be necessary to limit the outer layer surfaces to the component's contact patterns, feedthrough pads, and a short conductor trace for connection. This is sometimes referred to as "pads only"—with all circuit traces buried on internal layers of the PC boards.

COMPUTER-AIDED DESIGN AND VIA HOLES

On most computers, auto-routing for SMT requires a via pad and a hole on all layers of the multilayer board.

To simplify this procedure, a standard contact or footprint pattern for each type of device would include a line- or breakout-trace connection to the via pad. With this system, only the footprint appears on the surface, and all the signal traces are buried on inside layers of the board. With all circuits on inside layers, it is more practical to use .008, .006 inch or less trace width and air gap to interconnect the components.

SOLDER MASK ON PC BOARDS USING SMT

Specifications for solder mask and plating always promote controversy. Some companies would like to eliminate the solder mask altogether because of the extra cost and difficulty in registration on the small surface-mounted footprint. Others insist on the coating to seal the exposed laminate surface and insulate signal traces from possible danger from foreign particles or contamination. In either case, except for very small (3×4 inch) boards, the use of photo-imaged solder mask is recommended. See FIG. 8-14.

With a photo-image solder mask process, zero clearance from the footprint pattern is possible, but providing a clearance of up to .005 inch is usually more forgiving. Avoid a solder mask overlap or solder mask overlap residue on the contact area itself. At the same time, if too much

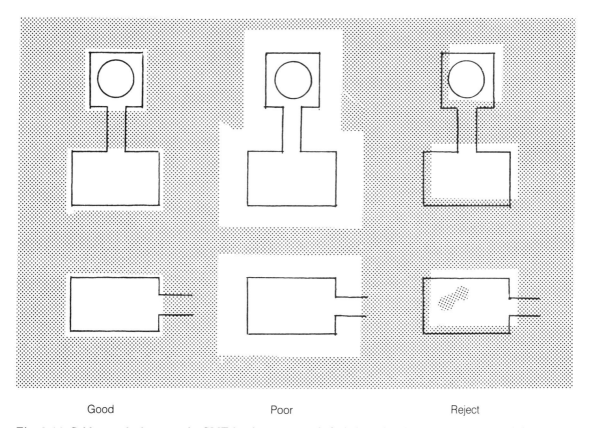

Good   Poor   Reject

Fig. 8-14. Solder mask clearance for SMT land patterns and via hole and pads must be kept at a minimum. An excessive opening for these features will allow solder migration, and solder mask residue on SMT contact areas is not acceptable.

of the signal trace is exposed, solder paste will migrate away from the contact area during the reflow-solder process.

The solder mask acts to contain the paste and to ensure that each component contact receives an equal amount of solder. Contact areas are usually connected to a via pad with a narrow trace. This isolation is necessary to allow for a dam of solder mask to stop solder from flowing to and down through the via hole.

A coating over the via hole is required if the via pad is too close to the contact area to provide for an adequate solder mask barrier. Some companies choose to cover all via pads with solder mask to eliminate any possible problems.

Note: Do not cover pads or vias needed for test probe contact by auto-test or bed-of-nails fixtures required for automatic testing.

The designer should discuss specific requirements with the board fabricator. There are excellent liquid and dry photo-image products on the market and this is a great opportunity to use the prototype phase of product development to evaluate the various candidates.

PLATING PROCESS FOR SMT

The most process-compatible plating required on the contact area is tin-lead. Tin-lead is a customary plating for etched copper circuits, although tin-nickel is frequently used. The advantage of the tin-nickel surface is the flatness and uniformity of the finished surface. Other fabricators, to achieve that flatness, will strip the tin-lead from the board after etching, leaving bare copper exposed over all traces. Solder mask is applied over the bare copper (SMOBC) and contacts are plated as a post-operation.

To be compatible with the surface-mounted devices and the solder and flux used, it is necessary to plate the contact areas, and any plated through-hole and pad mounting a leaded device, with tin/lead. With the bare copper board, solder mask is applied, exposing only the contacts and holes. Dipping this board into a tin-lead solder bath with hot air leveling adds the proper alloy to the contact areas for process compatibility. The same technique can be used on the nickel plated board. If solder mask is to be eliminated, electroplating of tin or tin-lead with a selective process must be used on contact areas and through-holes for mounting leaded components. In all cases, the tin-lead should be reflow-fused.

FABRICATION NOTES

The following printed wiring fabrication notes describe a 4-layer etched circuit board with PTH components and SMT on one side, photo-imaged solder mask over bare copper, tin-lead plating on holes, pads, and SMT contact areas. S/S legend is on one side.

Note, unless otherwise specified:

1. Material shall conform to IPC-ML-950B (FR4 type board) with 1 oz. copper on internal signal layers (if applicable) and 2 oz. copper after fabrication on external layers. Material must meet UL rating 94V-0. Final composite thickness will be .060 inch ± .003 inch.

2. Finished board shall meet requirements of IPC-A-600. Finished trace width and outside diameter of plated holes to be within .002 inch of image on photo tool master.

3. Reference: Photo tool artwork conforms to IPC-D-300-G, Class C.

4. Holes to be plated through in accordance with IPC-TC-500. Dimensions shown are for finished plated through-holes.

5. Edge connector fingers (if present) shall be plated with a minimum of .000030 inch of gold over .000200 inch of nickel plating material per MIL-STD-275.

6. Solder mask with photo-imaged liquid polymer or dry film on both sides of board in accordance with IPC-SM-840, Type B, Class 2 (over bare copper). Contact area of SMT components to be free of solder mask.

7. Solder coat on SMT contact patterns, holes, and pads, in accordance with IPC-D-320, Class 3, solder to be fused and leveled.

8. Screen legend with white non-conductive ink. Contact areas of SMT components to be free of ink.

9. Configuration of printed wiring (circuit) board not specifically dimensioned shall be controlled by photo tool artwork master.

10. Radius on inside corners and cutouts is acceptable. Maximum + radius of .060 inch is acceptable for trimming.

11. Board must meet all UL requirements. Supplier's UL identification to be affixed in an area free of copper, components, and legend. Boards will be tested for opens and shorts (bare board test) and confirmation of that test delivered with each lot.

12. No plating in tooling holes and other holes or areas, where specified.

BARE BOARD TEST

Bare board test fixtures used to verify the quality of the pin-through-hole type circuit board probe only the plated holes connecting sides or layers of the finished substrate. This system detects breaks or openings in trace paths, bad hole-plating and shorts.

In testing a board with surface-mounted components, the continuity of the plated holes is only part of the quality issue. To verify the circuit, the test probe contact must be made at the end of every *feature*. A feature is the footprint or/and pattern area that the SMT component lead-contact is attached to during the solder process.

These features, unlike the PTH boards, often have extremely close lead-to-lead spacing. It is common to have center-to-center spacing between leads of .050 inch on the small outline SOIC and PLCC ICs. The probes for testing these boards are individual spring-loaded contacts held in a custom-drilled polycarbonate panel fixture that aligns to the lead patterns of the devices, staggering pins on the rectangular footprint areas when required. With the addition of fine pitch devices, .0315 inch (0.8mm), .025 inch, .020 inch and smaller lead spacing, test fixtures have taken on a complexity new to the test process. Using test block concepts of fine wire probe clusters, a contact point can be made at each fixture. Test equipment manufacturers and even contract test houses are developing innovative techniques to provide the customer with reliable test data.

FABRICATION PROCESS FOR FLEXIBLE CIRCUITS

Flexible circuit wire interconnections have been used for numerous applications over the years. This concept of a very thin, rugged, and flexible interface medium, has proven beneficial, and when planned carefully, economical. The basic substrate materials in use today range from low-cost polyester to an extremely versatile polyimide that can withstand the high temperatures experienced with SMT assembly processing. Copper foils of various thicknesses can be laminated to the base material. These will accommodate chemically etched circuit patterns and outline shapes limited only by the designer's imagination.

Before a new flex circuit design is begun, a review of typical fabrication sequences and familiarity with recommended design guidelines will save costly errors, as well as unnecessary and expensive tooling charges.

The logical sequence of events for some typical (and some not so typical) developments in flexible circuits is:

1. Base material selection and copper foil lamination
2. Imaging and etching circuits on flexible materials
3. Pre-processing of coverlay (insulating) film
4. Coverlay lamination and piercing operations
5. Post plating options and die cutting techniques

BASE MATERIALS

The base material chosen to manufacture the flexible circuit must withstand a wide variety of process steps. The polyester material can tolerate process and environmental temperatures up to 200 degrees C. This relatively low-temperature material is acceptable for many interconnection requirements or low-stress applications. The thicknesses of lower cost polyester can range from .002 to .010 inch.

Polyimide base materials, although more expensive, are tougher than the polyester and will withstand temperatures beyond 370 degrees C. Polyimide base material thickness can be as thin as .001 inch or up to .005 inch thick, with excellent dimensional stability. Base material of .002 inch thickness or greater is preferred, however, due to the ease of processing and handling.

COPPER FOIL LAMINATION

Copper foil is press laminated with .001-inch acrylic adhesive to the base material. The foil is cut into manageable panel sizes of 12×12 inches or 12×24 inches, depending on the fabricator. Copper thickness ranges from one-half ounce to two ounces. The one ounce copper (.0014 inch thick) is used most often. Copper can be laminated to both surfaces of the base material, providing for two-sided circuits, or when necessary, laminated into multilayer substrates. These multilayer circuits, although less flexible, are processed in a similar fashion to the conventional rigid printed circuit board.

IMAGING AND ETCHING

A dry resist film is first laminated to the copper clad base material. Photo tool artwork of multiple images of the circuit is often prepared on glass plates. The composite is "sandwiched" between glass circuit image masters. The glass master is a far more precise medium than a typical film photo tool used in conventional PC board fabrication. After exposure of the light sensitive resist film on the copper, the exposed material is washed away, leaving a resist pattern over the proposed copper circuit. After curing the retained film image, the exposed copper area is chemically etched away. Methods of etching differ from one fabricator to another, but generally, high pressure spraying of the liquid etchant is used to remove the copper material. The copper traces and space between traces can be as little as .004 inch on one-half ounce and one ounce copper-clad material.

COVERLAY PREPROCESSING

Coverlay material is the same type of durable polyimide used for the base layer. The coverlay can be as thin as .001 or .002 inch and is applied to the base composite after etching the circuit pattern. All openings in the coverlay are pre-drilled, milled, or punched before lamination to the base circuit panel. These openings may provide access to connector patterns or the land pattern of the SMT components. Coverlay material thicker than .002 inch is possible, but not recommended for SMT applications. The components may not seat into the solder paste if the coverlay stock is too thick, causing a teetering characteristic and encouraging component shift, or *tombstoning*, when one end of the component rises off the substrate material.

COVERLAY LAMINATION AND PIERCING

The coverlay is "pin registered" to the finished base layer circuit and subjected again to the high pressure lamination process. On production processing of the circuit, a "hard tool" die is used to punch-pierce

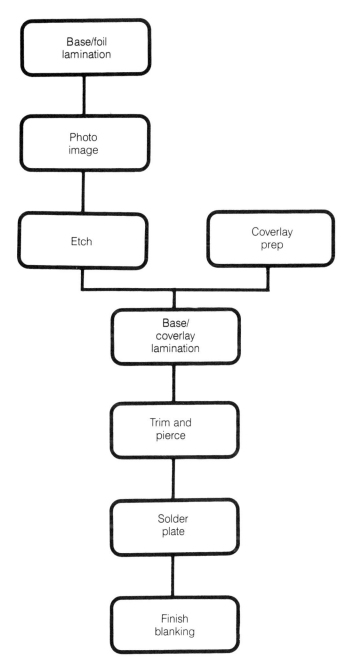

Fig. 8-15. The illustration describes a basic flexible circuit fabrication procedure, which will vary with complexity of the design and with the number of layers of conductor traces.

holes or shapes into each of the circuit images on the laminated panels. Prototype or short-run panels will use milling or drilling equipment to provide openings or holes. This method is more costly but saves the expense of "hard tooling" and lead time during the start-up phase of the product. FIGURE 8-15 illustrates a typical fabrication flow.

POST PLATING AND DIE CUTTING

Soldering leads to bare copper contact pads and holes is a common practice; however, tin-lead plating SMT contact locations is preferred. The tin-lead alloy is a more process-friendly medium for reflow-solder attachment of surface-mounted devices. To provide a more solderable surface to the flex circuit, electroplating or reflow plating of the tin-lead alloy can be used. A popular method of solder coating the SMT contact areas is the screen or stencil transfer of a 37/64 ratio tin-lead solder paste to the pattern. The solder paste is then fused to the copper area using a high-temperature reflow followed by a leveling process as shown in FIG. 8-16. Because of the many variables in this process, some variance in the finished plating thickness should be expected.

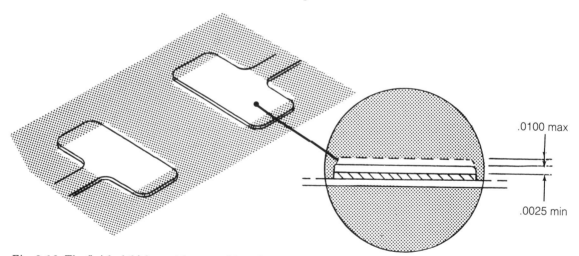

.0100 max

.0025 min

Fig. 8-16. The finished thickness tolerance of the solder on the contact areas is difficult to control with reflow-plating technology.

DIE CUTTING

To separate each circuit from the panel, the panels are passed through a blanking die designed to "punch-out" the finished profile of each flexible unit. A blanking die can be a relatively low-cost "steel rule" type fixture for lower volume and short-run products. The hardened tool steel, matched die systems can be developed for the mature long running parts. "Hard tooling" is more expensive but requires very little maintenance over the life of the product. For flexible circuits with secondary SMT assembly processes, the blanking procedure could be postponed until all assembly operations are complete. Limitations and concerns for this option are outlined in chapter 2.

HIGH-TECH MATERIALS
FOR MILITARY APPLICATIONS

Ceramic (alumina) materials have traditionally been used in military or extreme environmental applications. But the need for larger substrate size and the less-than-uniform shapes of today's electronic applications have made it necessary to explore alternative materials. Metal core laminated construction of circuit boards using relatively new materials is now being accepted for military and space applications.

A metal core (Copper Clad Invar/Polyimide-glass or Polyimide/Kevlar) laminated substrate is being offered as an alternative material for applications using leadless ceramic chip carriers. Previously, the choice was limited to ceramic substrates in order to match the component's TCE (Thermal Co-efficient of Expansion).

The metal core laminate material has been approved for many applications, and companies are specifying CCI in new products being developed. However, extensive testing, technical papers, and reports have not resulted in either military specifications or applicable guidelines to assist the engineer and PC designer in implementing this process.

In this next section, practical guidelines are presented for the PC board designer to assure the most producible finished product.

SPECIFYING COPPER-CLAD INVAR

In order to have a clear understanding of CCI fabrication guidelines, it is important to know the process steps. One method of adapting these more stable materials into your products is by laminating the finished PC boards to a layer of CCI to provide stability, ground plane, and a thermal transfer medium as shown in FIG. 8-17.

The most basic metal core laminated substrate would have one or two CCI layers and one single-sided copper-clad polyimide sheet laminated to both outer layers detailed in FIG. 8-18. Lamination would take

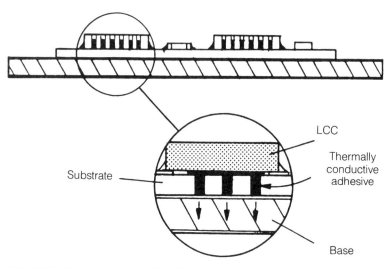

Fig. 8-17. Thermal compounds will transfer heat away from the component body and into the metal core layers or base of the substrate.

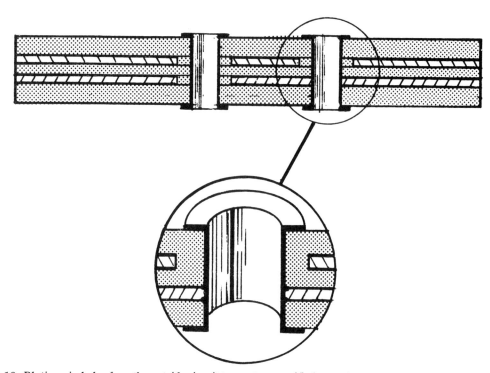

Fig. 8-18. Plating via holes from the outside circuit traces to a specific layer of a copper-clad invar core will require pre-drilling of clearance holes in other CCI layers before lamination.

place after clearance holes are drilled or etched in the CCI layers. This will prevent unwanted signal connection when holes are plated through.

Clearance holes must allow for insulation around via holes that are not to be connected with CCI layers. The laminating adhesive film will flow into the clearance holes in the CCI to provide the dielectric separation of copper plated via holes. After lamination, add all holes to substrate, including those that will connect to CCI layers.

The entire substrate panel is plated with one ounce of copper; and simultaneously, holes are copper-plated from one side to the other. Only holes not drilled for insulated clearance will connect to the internal layer, providing ground or plane connection. The circuit image is now plated to the outside surfaces and through the holes. This plating will act as a resist while exposed bare copper is etched away. If a tin-lead alloy is used as the plating medium, the etched panel is generally passed through a reflow process to fuse the tin-lead with the copper, leaving a bright finish.

A solder mask coating can be applied to both surfaces of the finished substrate. Contact areas and via pads should be free of this mask material, as noted for conventional PC boards.

The advantages of coating the substrate are: the solder paste used to attach the SMT components is contained on the land pattern, solder bridging is reduced or eliminated and, the bare laminate and conductors are covered, reducing absorption of moisture and contamination.

If space permits, reference designators and component outlines are screen printed on the board surface in epoxy ink. This aids in identifying each device as well as in identifying the polarity and orientation of SMDs.

The fabrication procedure will vary from complex applications that require the stability of a metal core laminated substrate. This variable depends on the number of layers required to interconnect the circuit. Surface-mounted components are often attached to both outside surfaces of the substrate.

It may be necessary to reduce the number of side-to-side plated through-holes to avoid excessive perforation of the metal core material. Many of the circuit interconnects can be made within the layer associated with each respective side. Only the via holes necessary to connect side one with side two and the power/ground CCI layers will be drilled during this step.

After the final drilling, copper plate the fully laminated substrate. Simultaneously, the copper plating connections through all via holes and

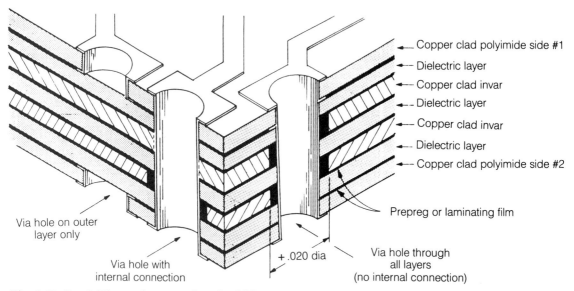

Fig. 8-19. *Pre-drilling and construction of a CCI core board will require a great deal of planning and interactive communication with the fabrication specialist.*

the PC board are processed as described earlier in the basic substrate fabrication examples.

The process requiring *blind vias*—holes that do not continue through all layers—will add to the circuit board fabrication cost. The continuity of the copper-clad invar layers, however, will be maximized by reducing the number of holes in the CCI. This maintains the continuous core layer, which is desirable to ensure the most thermally stable finished substrate (FIG. 8-19).

Most design rules recommended for surface-mounted technology on conventional epoxy glass substrates will apply to copper-clad invar-polyimide substrates. The assembly process control and solder selection may be more critical, but the components will be attached with reflow-solder technology in the same procedure as normally used in SMT assembly.

MATERIALS FOR COPPER-CLAD INVAR-POLYIMIDE SUBSTRATE FABRICATION

- Standard material thickness for polyimide-glass and polyimide-kevlar is .005, .006, .008, and .010 inch. Polyimide-Kevlar material presently costs four times as much as polyimide-glass. A careful selection process is recommended. Check your supplier to learn the standard sheet sizes available.

- In addition to the basic dielectric, choose the copper thickness desired:

> .0007 thick—½-oz Copper-Clad
> .0014 thick—1 oz Copper-Clad
> .0028 thick—2 oz Copper-Clad

Materials available are: bare; copper-clad on one side, copper-clad on two sides; or copper-clad on two sides, each having a different thickness. The latter is nonstandard and will be more costly.

- Prepreg, or laminating film is used to bind layers together and is furnished in a standard .0025 inch thickness. This material can be layered to build up to the final thickness specified for the finished board. Prepreg is supplied to the PC board fabricator in 38 inch-wide rolls.
- Copper-clad Invar material is available in several thicknesses—.006 inch is furnished in 24½ inch wide rolls. Materials that are .010, .020, .030, .050, and .060 inch thick are supplied in sheets.

Use standard material sizes whenever possible to keep costs under control; however, all of these materials are presently more expensive than conventional FR-4 epoxy glass laminates. The additional cost is far less than multilayer ceramic-substrate materials and mechanically the assembly will prove more durable.

RECOMMENDATIONS TO ENSURE SUCCESSFUL CCI FABRICATION

1. A symmetrical assembly that has equal stresses on both sides of the substrate at all times. The finished PC board should remain flat upon thermal cycling. Bonding the etched copper circuit polyimide layers on both sides of the copper-clad invar also eliminates assembly flexing problems during thermal cycling.
2. A conductor width of .010/.012 inch is preferred. .008 inch is the minimum conductor width on outside layers.
3. Total board thickness range is .062 to .100 inch with a maximum of .125 inch.

4. Board thickness tolerance is 10 percent of nominal or .007 inch, whichever is greater.
5. A .008-inch thickness is preferred for the dielectric between conductive layers. The dielectric layer should not be less than .0035 inch.
6. An air gap of .010/.012 inch between conductors is preferred, with a minimum of .005 inch.
7. Finished board thickness to plated through-hole size ratios: 3:1 is preferred; 4:1 is maximum.
8. Plated through-hole diameter to finished board thickness should be:

    .062 inch thickness or less: .020/.025 inch hole dia.
    .075 inch thickness or less: .025/.035 inch hole dia.
    .100 inch thickness or less: .035/.055 inch hole dia.

    Hole sizes .015 or .018 inch diameter, if used, are to be used on outer layers only.
9. Annular plating around finished hole is .007/.010 inch—minimum of .005 inch is to be avoided.
10. Conductor clearance to edge of board: internal layer of .100 inch is preferred, .031 inch min. External layer of .100 inch minimum.

Most of these recommendations are applicable to conventional PC board fabrication as well. Specifying standard materials will ensure greater producibility, since this allows the fabricator to use off-the-shelf materials. Overall thickness can be specified, including plating, but do not specify spacing between the layers created by the prepreg or core thickness. If you must specify core and/or prepreg thickness when spacing is critical, give a loose tolerance on the overall thickness for maximum fabrication efficiency and yield.

9
SMT
Assembly Process

THE ASSEMBLY EQUIPMENT AND PROCESSES FOR SMT are quite different from the pin or leaded through-hole component method of assembly. Initially, a bench-top SMT assembly station will suffice for a company with small-volume assemblies. As products move into higher volume, robotic and automatic assembly equipment becomes more practical.

The manufacturing and assembly process in SMT today involves a large percentage of circuit assemblies that use both leaded IC components and surface-mounted devices on the same PC board. FIGURE 9-1 describes a flow diagram of a medium- to high-volume assembly line for mixed-technology PC boards.

EQUIPMENT REQUIREMENT FOR SMT

Setting up an in-house SMT line is frequently the task of those new to the technology. Manufacturing services are often used as a line parallel to the customer's own operation. This builds in a safety factor to allow for reaction to sudden surges in demand. To plan properly for SMT equipment needs, review your current requirement and future projections for:

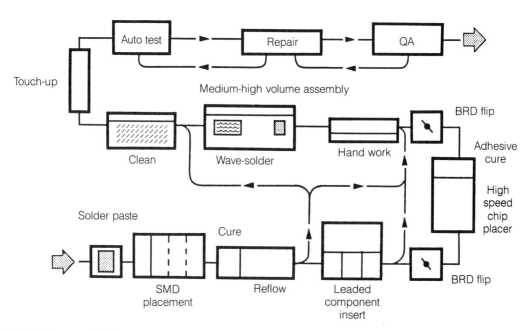

Fig. 9-1. The assembly flow of the SMT product can include several process steps. Each process step is audited to maintain the established quality yield.

142

- Component types and quantity per assembly,
- Number of different assemblies to be processed, and
- Current and projected volume requirement per assembly type.

The contract manufacturer is usually prepared for a very wide variety of assembly types and component mixes. More successful organizations have the resources to respond to your success. In selecting a manufacturing partner, make sure they have the capital, equipment and manpower to assure expedient ramp-up when necessary.

PLANNING THE SMT ASSEMBLY PROCESS

The flow of your product through a typical assembly operation requires planning. Experienced planners and manufacturing engineers can ensure the best use of equipment and human resources, assuming all component parts are in the pipeline and these components are furnished in the correct packaging required for the automated SMT assembly systems. Tape-and-reel packaging is specified for the smaller or passive components and ICs are available in tube magazines, tape-and-reel for the high-volume applications, or waffle pack trays in the case of Quad Flat Pack (QFP) devices. Communication, coordination, and cooperation between the design engineers, buyers and manufacturing engineers in this planning stage is vital!

CASE STUDY PRODUCT DESCRIPTION

The following is a description of an SMT assembly with a small number of leaded parts that will be hand-installed as a post process:

1-	PC board (3.5×11.2 inch)	73-	Passive components
24-	Small outline ICs	12-	J-lead PLCC ICs
2-	120 lead QFP ICs	2-	40 lead DIP ICs
2-	PTH connectors	4-	Radial lead devices

This assembly provides for all components to be mounted on one surface. Solder paste is applied to the SMT contact patterns with an etched stencil fixture.

Surface-mounted component parts are placed into the solder by a combination of robotic systems followed by an infrared (IR) reflow process and a solvent cleaning station. We will assume that the leaded parts will be installed and soldered in a wave system, or by hand.

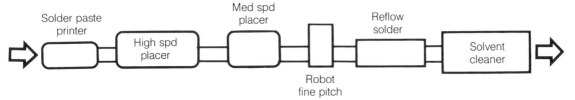

Basic automated SMT assembly line

Fig. 9-2. The basic SMT assembly system for reflow-solder technology may include one or more placement machines, depending on device type and quantity.

FIGURE 9-2 illustrates the flow of a board assembly through a series of processes. The combination of compatible systems is matched to the specific volume and component type mix of the product. Unlike a dedicated line of one product, a flexible combination of systems for various applications can be quickly set up for a multitude of product types and quantities.

To expand on each element of the surface-mounted assembly process, the following text describes the materials and equipment in this sequence:

1. solders for component attachment
2. using solder pastes with SMT
3. solder application
4. solder screens and stencils
5. assembly methods
6. assembly options for SMT
7. curing solder paste
8. reflow process
9. cleaning after reflow soldering
10. providing for automatic vision alignment
11. preproduction design review

SOLDERS FOR COMPONENT ATTACHMENT

Solder, an alloy made principally of tin and lead, provides reliable electrical connections and mechanically strong joints, when and if the correct alloy is selected. Other metals such as antimony, silver, cadmium, indium, and bismuth are alloyed with tin and lead to control certain physical and mechanical properties of the alloy, e.g., melting range, tensile and shear strength, hardness, and even corrosion resistance.

Tin-lead or "soft" solder alloys are the most widely used in electronic applications, since their low melting temperatures make them ideal for rapid joining of most metals by conventional heating methods.

Take care in specifying the proper alloy for each soldering job since each alloy has unique properties. When referring to tin-lead alloys, tin is customarily listed first. For example, 60/40 refers to 60 percent tin, 40 percent lead, by weight.

General purpose solders include 40/60 and 50/50, which are typically used for plumbing and sheet-metal as well as for electrical applications. Where minimum heat must be used during formation of the solder joint, as in surface-mounted assemblies with heat-sensitive components and materials, higher tin-content alloys are required, such as 60/40 or 63/37.

Alloys of tin-lead with a small percentage of silver (63/35/2) are used to reduce the leaching of silver from silver alloy end terminations of some passive components. These types of alloys are also ideal for soldering to thick-film silver alloy coatings on ceramic hybrid circuits.

Bismuth-containing solders, which are frequently used as fusible alloys, may be used in applications where the soldering temperature must be below 381° F (183° C). Indium alloys, also with low temperature melting ranges, are very ductile and are therefore suitable for joining metals with greatly different coefficients of thermal expansion.

USING SOLDER PASTES WITH SMT

Solder pastes are homogeneous mixtures of a paste-flux and fine powder solder alloy. The physical and chemical characteristics of the material can be matched precisely to the solder joint requirements, e.g., the method of placement used and required definition, in-process conditions, solder-reflow method used, and cleaning requirements.

Since all the ingredients required to successfully place and solder the components are contained in the paste, it is an ideal material for automated assembly of simple or complex mechanical, electrical, and electronic systems. Reliable solder joints can be repetitively produced in conventional applications. For difficult assembly applications, solder paste may provide the only practical method of inexpensive solder attachment.

Electronic grade solder pastes are manufactured to meet the critical requirements of electronic component assembly. The composition of the pastes can vary with individual requirements. A wide variety of compositions may be specified from suppliers, which comply with recognized standards, like those of the ASTM (American Society for Testing and Materials).

Flux used in solder paste consists of the following primary components: (1) solvent or vehicle; (2) rosin/resin or synthetic (solid); (3) activator; (4) viscosity-control additive.

1. Solvent selection is based on compatibility and ability to dissolve high concentrations of solids. Other important properties are: slow evaporation at room temperature, low moisture absorption, high flash point, and compatibility with supplemental activators and viscosity modifiers.
2. The solids used in solder pastes are selected for their characteristics. These include: cleaning of surfaces to be soldered, oxidation protection to the solder powder and solder joint during heating, binder and surface protection after curing, stability at soldering temperatures, and ability to be completely removed by conventional cleaning solvents.
3. Activation levels of flux include: non-activated (R), mildly activated (RMA), activated (RA), and super-activated (RSA).
4. Organic thickeners are added to flux systems to alter physical properties as required for typical applications. Thickness is selected based on methods of dispensing—screening, stencil printing, or other.

Solder-flux residues must be removed from the board surface after reflow. Both organic and inorganic activators are available. Organic systems may be preferred for ease of cleaning and environmental acceptability and safety.

SOLDER APPLICATION

Solder pastes are dispensed in several ways, including syringes and pressure-fed reservoirs of guns. Large or small amounts of paste can be dispensed this way. Single dots or strips may be controlled with timing devices or robotic applicators. Another application method is casting component contacts with solder paste before attachment to the PC board footprint.

SOLDER SCREENS

Screen transfer application is a widely used method of depositing solder paste to the SMT footprint pattern. The paste is relatively easy to apply, and precision is controlled.

Solder pastes can be accurately transferred onto a surface, as shown in FIG. 9-3, by a manual or automatic operation, using screens ranging from 80 to 200 mesh. Mesh size depends on the definition required. Paste thickness is controlled by screen size, emulsion, or mask thickness. A squeegee is normally employed to force paste through openings in the screen that contact-print the areas where paste is to be applied. All other areas are sealed.

Transferring solder paste with a stencil will improve accuracy. The stencil material, usually brass or stainless steel, can be specified in various thicknesses. The surface-mounted component land pattern is chemically etched through the material, leaving a precise opening for solder paste transfer. Thickness of the paste is determined by the gauge of the stencil material. For SMT assemblies with fine pitch QFP devices, the stencil is superior to the screen. Multilevel stencils are possible as well. See FIG. 9-4.

The process engineer will have the option of depositing .006 inch-thick solder paste on the fine pitch IC land pattern, while maintaining a .008-inch thickness of the alloy for the larger surface device type. Adja-

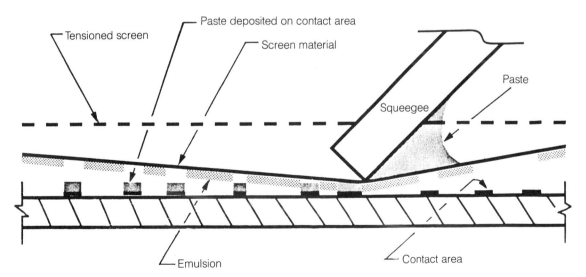

Fig. 9-3. Depositing the solder paste to the contact land patterns of the substrate is efficiently accomplished with screen or stencil fixtures.

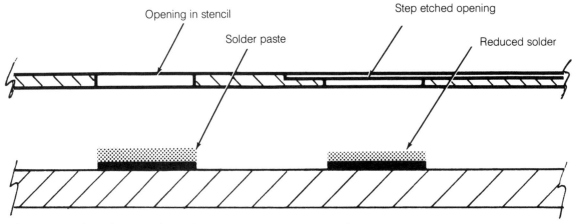

Fig. 9-4. Applying solder paste in varying thicknesses at select areas of the substrate can be accomplished with selective etching of the stencil material.

cent component clearance to the fine pitch ICs must allow for the squeegee to conform to the recessed area. Consistent solder paste resolution can only be maintained if the squeegee is in full contact with the stencil surface during the transfer sequence.

ASSEMBLY METHODS

The operation of placing surface-mounted devices onto a designated footprint or contact pattern is referred to as pick-and-place. For low-volume assembly or prototypes, hand placement with tweezers or vacuum pickup tools is adequate. As component count and density increases or assembly volume grows, the use of automation for this task becomes more practical. Speed, accuracy, and component versatility will vary from one manufacturer to another. Factors that must be addressed in choosing equipment for a factory are:

- maximum substrate or board size and placement area,
- volume projected for the product,
- component placements per hour,
- accuracy of placement and repeatability,
- part mix, number of component stations, and
- type of components and size limits.

After evaluating the product and component mix, it may be practical to have several different types of systems in line to efficiently process the assembly.

ASSEMBLY OPTIONS FOR SMT

Type One

Type One's a one- or two-sided PC board, as shown in FIG. 9-5. Surface-mounted components are attached on one side to contact or footprint patterns etched into the circuit. The assembly sequence starts with applying screen-printing solder paste onto specific patterns on the board surface. The components are placed onto the PC board surface, aligned with the footprint pattern, and cured. The curing chamber raises the temperature of the assembly to extract any solvent or moisture still present from the paste or the board itself.

After curing, the assembly is transferred to a higher temperature chamber until the solder paste is converted into a liquid and reflows the solder, creating the electrical and mechanical bond between the component and the PC board.

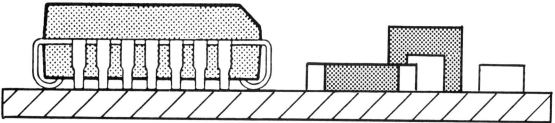

Fig. 9-5. The most basic Type-One SMT assembly will require only one solder process.

Type Two

Type Two is a two-sided PC board with surface-mounted components mounted on both sides, as shown FIG. 9-6. The assembly sequence would first apply solder paste to the contact or footprint pattern on side one of the PC board. The surface-mounted devices are placed into the paste material and cured and reflowed in the same manner as a Type One assembly.

Side two can be assembled in the same way, or the components can be attached with adhesive epoxy and wave solder. Solvent cleaning usually follows each assembly phase to prevent flux residue from hardening.

Type Three

Type Three employs leaded or through-hole components mounted on side one, and surface-mounted parts attached to side two with a UV

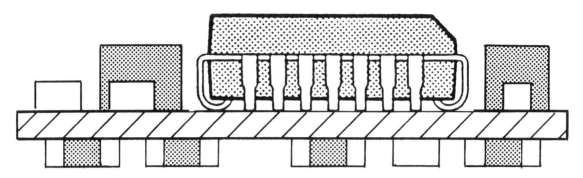

Fig. 9-6. Type-2 assemblies require a second solder process. The second process may use wave solder technology.

cured epoxy, and wave soldered simultaneously as in FIG. 9-7. The leaded components might include conventional DIP and SIP packages, axial-lead resistors, diodes, capacitors and jumpers or radial-lead devices.

On side two, the surface-mounted parts would incorporate chip resistors and capacitors, SOT and SO packages, diodes, etc. Tantalum capacitors in SMT packages may not always be suitable for side two mounting because of the higher profile of the component. Wave-soldering of the PLCC is also impractical because of its high profile and lead configuration.

The Type Three assembly was the very first approach to adapting SMT to high-volume production of PC boards. The economic advantage is twofold: greater component density, and utilization of common wave-solder equipment.

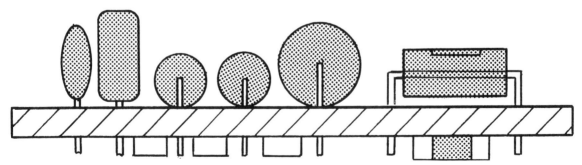

Fig. 9-7. The Type-3 assembly requires mixed technology, with leaded parts on one side and SMT devices attached on the other with epoxy for wave solder.

Type Four

Type Four assembly, as shown in FIG. 9-8, attaches lead- and surface-mounted devices on side one with additional surface-mounted components attached to side two.

The goal of component specialists is to choose as many surface-mounted alternatives to the leaded predecessor as possible. With few exceptions, most through-hole devices are furnished in surface-mounted packages.

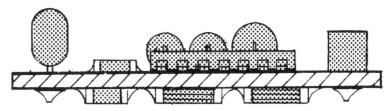

Fig. 9-8. The Type-4 SMT assembly, the most complex type, mixes surface-mounted devices on both surfaces of the substrate with leaded-PTH devices on one of the sides.

CURING SOLDER PASTE

Because of its tackiness, wet solder paste acts as a retaining material for the miniature SMT components. The retention of this tackiness must be considered in planning the production assembly. After placement, the curing cycle prepares the board for the solder reflow. During the drying or curing process, the normal sequence of events is:

- the flux solvent starts to evaporate,
- viscosity starts to decrease, depending on solids in flux, percentage of metal, and thickness,
- solvent is completely volatilized,
- water-white rosin or heat-stabilized rosin/resin starts to melt and forms a protective envelope.

When curing is complete, components are held in place and the absorption of environmental moisture is eliminated. An optimum drying system is one that slowly heats the assembly from the ambient to about 85 degrees C. If some spread is acceptable, rapid drying using infrared heating is effective.

REFLOW PROCESS

Solder methods for reflow include: 1) ovens and induction heating; 2) infrared (IR); 3) conveyorized hot panel and; 4) hot air and vapor phase.

1. A circulated-air oven or electrically heated convection furnace, with or without inert gas, can be used effectively to reflow solder paste. Both a batch-type oven and multi-zone conveyorized-belt furnace can be used.

2. Infrared belt furnaces, with two or three heat zones and top- and bottom-heating, will provide rapid reflow. Absorption of IR energy varies with materials used. For example, the organic components of solder paste are excellent absorbers of IR, while gold or aluminum and previously reflowed solder are good reflectors. The use of focused- or non-focused IR is an efficient method of reflow soldering individual components. Laser soldering with a microprocessor controlled IR system is extremely fast; however, the paste must first be dried due to the rapid heating cycle of this method.

3. An electric hot panel, or a series of hot panels, with a belt to transport parts is an effective method of heating parts by conduction.

4. Vapor-phase or condensation-soldering offers precise temperature control. The entire assembly is heated to the temperature of the vapor of a chemical held at its boiling point (FIG. 9-9). Both batch and in-line vapor-phase reflow systems are available.

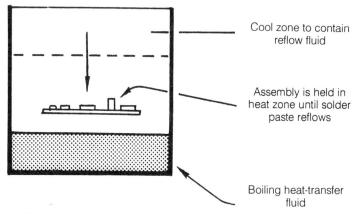

Fig. 9-9. Vapor-phase solder-reflow methods are ideally suited to those substrates that have a solid metal core, or assemblies with very high component density.

CLEANING AFTER REFLOW-SOLDERING

When cleaning to remove the flux residue, a solvent-type cleaner should be used for two reasons: 1) the flux, emulsifiers, and thickeners in the solder paste do not readily dissolve in an aqueous solution; 2) the space under some of the surface-mounted components is approximately .003 inch. The water molecules are not small enough to get under the components and flush out contaminants. (See FIG. 9-10.) The cleaning equipment selected is dependent on volume and company specifications for assembly cleanliness.

The cleaning method to remove residues from rosin, resin, or related synthetic-solid-based solder paste must be matched to components used in the soldered assembly. Fluorocarbon-type solvents with added polar solvent have excellent penetration characteristics and may be used in different cleaning cycles, including ultrasonic systems.

Bipolar chlorinated solvents are also effective, when followed by water cleaning. Residues from water-soluble flux systems should be removed with agitated or boiling water, followed by de-ionized water rinses and forced-air drying.

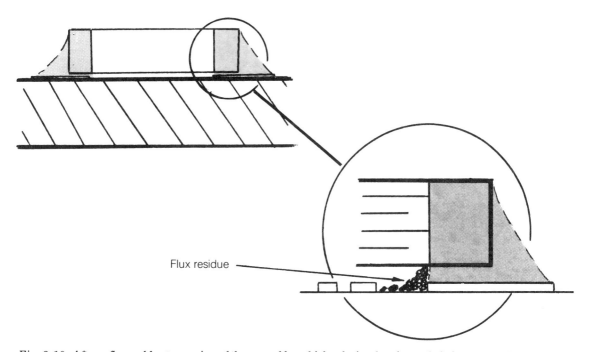

Flux residue

Fig. 9-10. After reflow-solder processing of the assembly, a high velocity cleaning cycle is incorporated to remove all flux residue and foreign particles.

PROVIDING FOR AUTOMATIC VISION ALIGNMENT

The fiducial target method of alignment will be used for automatic solder stencil and assembly operations. Three fiducial targets are required on each panel to permit automatic alignment. The fiducial location can be outside the assembly area of a panel or on each individual unit. To provide for automatic placement of fine pitch and TAB devices, the addition of two targets within the contact pattern is advised. (See FIG. 9-11.)

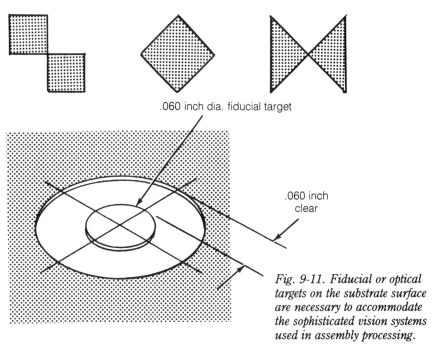

.060 inch dia. fiducial target

.060 inch clear

Fig. 9-11. Fiducial or optical targets on the substrate surface are necessary to accommodate the sophisticated vision systems used in assembly processing.

One fiducial target is positioned at the center point, the second at the outside corner of the component. Fine pitch usually defines a high pin-count IC with center-to-center lead spacing of .031 inch or less. Many devices now being used for custom and semi-custom applications are using .025-, .020-, and .015-inch spacing. The trend moving toward even closer spacing requirements will continue until the limit of chemical etch technology and the placement accuracy of component assembly systems is reached.

The assembly sequence would start with: 1) solder paste application to the surface-mounted component's contact patterns, 2) mounting the components in paste, and 3) curing. The PC board passes through a

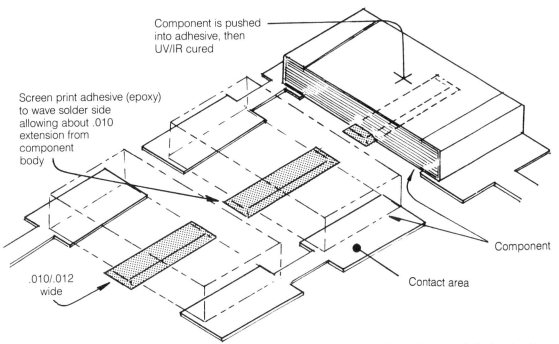

Component is pushed
into adhesive, then
UV/IR cured

Screen print adhesive (epoxy)
to wave solder side
allowing about .010
extension from
component
body

.010/.012
wide

Component

Contact area

Fig. 9-12. Stencil application of epoxy for component attachment on the wave-solder side is practical when lead parts are added as a secondary operation.

high-temperature chamber that melts or reflows the solder paste, leaving a finished solder connection on each contact of the surface-mounted device. The assembly is then cleaned, ready for the next phase.

To mount the remaining surface-mounted components on side two, an adhesive epoxy is applied to retain the device, as in FIG. 9-12, and again cured to harden in position. With the parts now bonded to side two, the board is flipped back to side one and all remaining component leads are hand inserted through plated holes in the board.

Caution: If leaded parts are to be machine assembled, the epoxy attachment of side two SMT components will follow insertion and crimping operations. Following the above sequence, the assembly is passed through the wave-solder process and cleaned once again before inspection.

PREPRODUCTION DESIGN REVIEW

In the rush to get a product to market, engineers and designers do not always adhere to volume manufacturing guidelines. To prepare the product for volume assembly, it is sometimes necessary to make adjustments in the design. In some cases, component spacing or footprint patterns require minor changes to be compatible with automatic systems or processes. Provide adequate time to evaluate the assembly for "manufacturability." The following chapter will define guidelines that will assist in evaluation of the SMT assembly.

10

Design Evaluation for Volume Production

AS PART OF THE MANUFACTURING PLANNING CYCLE, A task group is first assembled to evaluate the producibility of an SMT product. The review committee would include representatives of several disciplines—design, manufacturing and process engineering, test, quality, and materials. The review will provide engineering data in order to identify potential problems which manifest themselves as additional product costs. The list of concerns might not necessarily be assembly problems or, more specifically, oversights; however, they should be identified because many of them will incur additional manufacturing costs.

Concerns fall into several categories: documentation errors, potential interference problems for automatic placement of SMDs, component pad sizes, and printed circuit board interconnect techniques.

Before the critique of the surface-mounted assembly, prepare a checklist of primary concerns that may affect each phase of the assembly process.

SUBSTRATE QUALITY

A sample checklist would be similar to the following:

_____ Board flatness and appearance
_____ Tooling holes for machine handling
_____ Photo-imaged solder mask
_____ Solder mask over bare copper
_____ SMT contacts free of mask
_____ Screen legend is sharp and readable
_____ Legend does not overlap pads or SMT contacts
_____ Trace width and air gap
_____ Hole size to pad ratio
_____ Hole registration/breakout
_____ Plating quality and uniformity
_____ Bare board test certification
_____ Verify dimensional accuracy

Inspect the substrate to the specified IPC quality level established by the design engineer. Following bare board inspection, apply solder paste to the SMD land patterns on a sample of non-populated circuit boards. A solder sample should be furnished with each board lot or date code. Reflow the solder paste and clean. The solder quality can be evaluated as well as the substrate's physical reaction to the temperature cycles of reflow systems. One or more thermal couple probes can be attached on the substrate surface to establish basic temperature profiles for the reflow system. These profiles will assist in establishing temperature and conveyer speed during reflow of the populated assembly.

Before the board is populated with components, inspect the quality of the plated via and lead holes. Cracks or voids in the plating could be a sign of more serious fabrication problems. The quality and finished product yield of the surface-mounted assembly is directly related to the quality of the substrate.

ASSEMBLY DOCUMENTATION

Assembly details for surface-mounted circuit boards must define the location and orientation of all component parts. This will include the component outline and reference designation for each device, including connectors and hardware. An orientation identification mark on the substrate surface must be provided for ICs, diodes and all polarized devices. Pin one of the IC or connector, as an example, will be defined by a small round shape on the screened legend as detailed in FIG. 10-1. Graphic shapes representing component bodies are often added to the legend master to clarify position on the board surface.

When preparing assembly documentation, provide a composite of the screen legend graphics over a halftone image of the circuit. Outlines of components not furnished on the legend master must be added to the drawing manually. If CAD data can be generated to plot a clear assembly

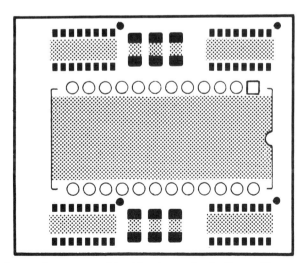

Fig. 10-1. Screened component identification and polarity markings must be clearly legible after the device is in position.

guide, the composite drawing can be omitted. The final assembly detail must include auxiliary views when necessary, to clarify mounting of unique components and notes to address special requirements. The assembly detail must furnish enough information for processing and inspection of the final configuration.

The bill of material (BOM) must include component description and type, manufacturer's part number, reference designator and quantity. Data sheets or specifications must be furnished for any components classified as special. An approved vendor list (AVL) should also be included as a controlled part of the documentation. Two or more sources should be provided for each part in the BOM when possible.

Circuit board fabrication drawings must include material, physical detail, and specific notes describing the finished product. Holes are identified and finished size and tolerance clearly defined. Multilayer substrates will require a cross-section, as shown in FIG. 10-2, detailing circuit layers and finished thickness. For more information on board fabrication and material options refer to chapter 8.

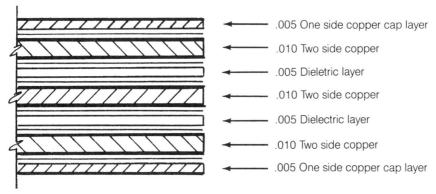

.005 One side copper cap layer

.010 Two side copper

.005 Dieletric layer

.010 Two side copper

.005 Dielectric layer

.010 Two side copper

.005 One side copper cap layer

Fig. 10-2. A cross-sectional view of the multilayer substrate will identify the copper-clad weight, dielectric layers, and finished thickness specification.

Following the documentation review, inspect photo-tool artwork that is used to manufacture the circuit board. Measure component land patterns and compare data to the device specifications. The wrong calculation of contact spacing for a fine pitch QFP, for instance, will cause extensive hand rework after reflow solder. Inspect clearances between traces and via pads on internal layers. The conductor width and air gap is often far less than the circuit patterns on the board's outer surface. Any clearance less than .006 inch should be avoided, and the same is true for trace width. Board fabrication technology is available for extremely fine line requirements, but this will increase costs.

ASSEMBLY EVALUATION

Following attachment and reflow soldering of surface-mounted devices, inspect the solder fillet on each component contact. Study the alignment of the passive components and the finished registration of multi-lead IC devices. Troubleshooting assembly problems will require a close look at cause and effect. First, identify and itemize the defect to determine if quality issues are influenced by board design, component quality, process control, or a combination of each. When the reflow-soldered surface-mounted phase of the assembly sequence has been evaluated, attach the remainder of the components to be wave soldered. The post wave-solder inspection should be just as critical as the reflow. Signs of solder bridging to adjacent leads or epoxy attached SMT devices should be noted. Clearance between component bodies and spacing from leaded devices must be adequate to allow a barrier of solder mask between solder contacts. Examples of problem areas are shown in FIG. 10-3.

The following checklist will act as a guide during evaluation of the assembly:

_____ Solder mask separation of land pattern and via pad

_____ Correct solder mask clearance around land patterns

_____ Solder mask not present on SMT land pattern surface

_____ Passive component spacing of .030 inch min., side to side

_____ Via hole and pad is not located under body of small chips

_____ Component-to-board edge clearance for machine handling

_____ Space between components for visual solder inspection

_____ Orientation of parts clearly identified

_____ Components aligned after reflow-solder process

_____ Correct IC lead registration and spacing

_____ Adequate solder fillet for inspection (and touchup)

_____ Lead spacing on PTH devices correct for machine insertion

_____ Non-plated tooling holes provided for machine fixtures

_____ Three fiducial (optical) targets for vision alignment

TRACE-TO-CONTACT GUIDELINES

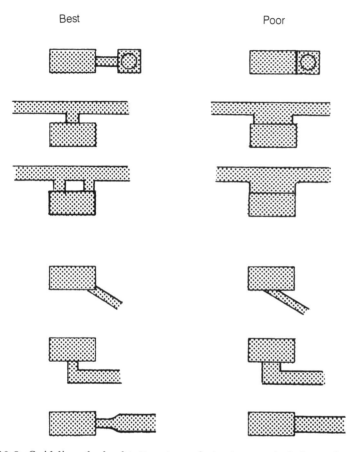

Fig. 10-3. Guidelines for land pattern-to-conductor trace and via-to-conductor trace interconnections have been well established. Adhering to these guidelines will ensure a "process friendly" design.

TESTABILITY

One of the most critical elements in circuit board design is a provision for automating testing of the final assembly. If test probe contact patterns are not accessible at each common junction of the circuit network, measuring or exercising the components individually in the circuit will not be possible. The location of the malfunctioning device is difficult and costly with bench top analyzing equipment. By using In-Circuit-Test systems, each part value can be measured, IC device functions can be checked, and the entire assembly screened for solder opens and shorts.

When reviewing the assembly for testability, verify the following requirements:

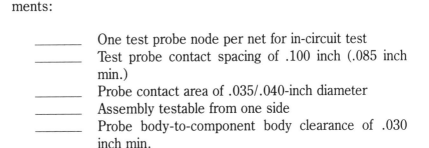

    _____ One test probe node per net for in-circuit test
    _____ Test probe contact spacing of .100 inch (.085 inch min.)
    _____ Probe contact area of .035/.040-inch diameter
    _____ Assembly testable from one side
    _____ Probe body-to-component body clearance of .030 inch min.

Do not rely on the component lead or land pattern for test probe contact. The pressure of the spring loaded probe pin will add pressure to a substandard solder connection. When the test pin contact is released, the component lead is allowed to open and electrical continuity broken. Probe contact to an irregular surface can also cause damage. Test probes are damaged when distorted and further damage will occur when the sharp alloy probe tip makes physical contact with component bodies.

Miniature test probe contacts are available for restricted assembly applications. These probes are designed for center-to-center contact spacing of .050 inch. The probes are more expensive and less durable than the conventional .100-inch spaced probe type.

GENERAL DESIGN IMPROVEMENTS

Following the recommendations, outlined and illustrated in the preceding chapters, will prepare the designer with the fundamentals necessary to produce a successful SMT assembly. In general, a few subtle elements of good design practice should be reviewed:

- Use process proven SMT land pattern design;
- Retain consistent orientation on components when possible;
- Mount polarized devices in the same direction;
- Allow reasonable component density and even distribution;
- Select SMT devices in standard configurations;
- Choose parts that will have multiple sources;
- Design the circuit board to control excessive costs.

The SMT design guidelines furnished in this book are to be used as tools to assist in developing the most successful products possible. In addition to proven design related factors, components and processes are being refined constantly from sources all over the world.

New components developed for application specific integrated circuits (ASIC) have increased lead density, which will require the development of unique land patterns, interconnection methods, and improved solder technology.

Although mature, surface-mounted technology continues to evolve, the design and assembly processes are interrelated far more closely than the PTH products of past years. Design engineers, CAD specialists and manufacturing disciplines must work together in planning a product for SMT.

More than an evolution, surface-mounted technology is a revolution in the ever expanding electronic manufacturing industry.

Glossary

0805 Basic physical size code for passive devices, .080 inch long by .050 inch wide.

1206 Basic physical size code for passive devices, .12 inch long by .06 inch wide.

1210 Basic physical size code for passive devices, .12 inch long by .10 inch wide.

1812 Basic physical size code for passive devices, .18 inch long by .12 inch wide.

2225 Basic physical size code for passive devices, .22 inch long by .25 inch long.

adhesive Liquid or film compound for attaching materials to one another.

air gap Clearance between conductive circuit traces on a substrate layer.

alumina substrate Ceramic base material used for additive conductive circuit traces.

annular ring Width of conductive material around a via or hole provided in the substrate surface or layer.

artwork Preparation of working film used in the fabrication of circuit substrates.

ASIC Application specific integrated circuits.

BOM Bill of material or list of parts for an assembly.

bridge networks A set of diode devices interconnected for a specific electrical function.

bridging The connecting of two or more circuit points with solder alloy.

C-pack   Commercial plastic integrated circuit package with "butt mounted" lead contacts on four sides.

CAD   Computer-aided design.

capacitor   A passive electron storing device used in interfacing electronic circuit functions.

carrier contacts   Contact leads for attaching leadless IC devices to a substrate.

carrier pallets   A partitioned tray used to store or transport electrical devices.

castilation   The amount of alloy that has been deposited on device leads or contacts after solder processing.

ceramic capacitors   A monolithic layered dielectric and alloy film device.

cermet   Material used in the manufacture of resistive devices.

chip resistors   A resistive film element applied to a thin ceramic substrate with each element end attached to an alloy end-cap terminator.

component contact   The lead or area provided on a device for electrical and mechanical attachment to the circuit conductors of a substrate.

compression contact connector   A connector designed to electrically mate to contact areas of two parallel substrate surfaces.

contact footprint   See land pattern.

contact geometry   The physical shape of the land pattern provided to electrically and mechanically attach electronic devices to the substrate.

contact pins   Leads of an electronic device, connector or module used for electrical interface.

copper foil   Thin copper alloy sheet material bonded to a rigid or flexible substrate for chemically etching of circuit conductors.

coverlay   The insulating material or film laminated over each outside surface on a flexible substrate having etched circuit conductors.

curing chamber   An oven or enclosure designed to cure epoxies or other materials at an elevated temperature.

desolder   The removal of a device from the substrate by extracting solder from the device leads and land pattern.

device   Electronic component.

die-cut   A tool designed to cut or punch a substrate or other flat material into a predetermined and uniform finished shape.

dielectric   An electrically insulating material used to separate conductive layers of a circuit or device.

diode array   See bridge array.

diodes   Device used to transform low current ac to dc; functions as a one way valve for conducting voltage.

DIP   Dual in-line pin device or component having two rows of leaded pins for terminating to circuit conductors through holes in a substrate.

discrete components   Nonactive passive devices.

DRAM Dynamic read/write memory device used to store and maintain electrically coded information.

dry film A dielectric layer of material laminated to the surface and covering selective areas of a substrate.

dual bed of nails A test fixture used to electrically test bare and assembled circuit substrates on two sides simultaneously with probe contacts.

dual diode A single device package housing two diodes.

edge mount contacts Contact pins or leads attached to the edge of a substrate or module assembly.

EIA Electronic Industry Association.

EIAJ Electronic Industry Association of Japan.

end-cap termination Electrical and mechanical contact area of a passive device.

epoxy Adhesive material used to attach devices to the substrate surface or other mechanical attachment of two or more parts.

epoxy cure The hardening of the epoxy material through the exposure to ultraviolet light and/or elevated temperature.

epoxy glass A composite of epoxy resin and glass fiber fillers.

epoxy kevlar A composite of epoxy resin and kevlar fiber fillers.

epoxy quartz A composite of epoxy resin and quartz fiber fillers.

feedthrough hole A plated hole to connect two or more circuit layers of a substrate.

FET Field Effect Transistor.

fiducial targets A shape or pattern retained in two or more locations on the substrate surface for assembly systems using vision alignment.

fillet junctions Electrical and mechanical solder connection to the contacts of a device.

fine pitch Integrated circuit devices with center-to-center lead spacing of .032 inch or less.

fixtures holding plates, clamps, solder paste stencils, and other hardware used to process a circuit assembly.

flat cable Insulated multiple paths of wire or conductive material routed in parallel.

flexible circuits Conductive circuit etched on one or more surfaces of a flexible substrate base.

flux Active chemical used to promote intermetalic bonding of solder to the substrate circuit and device lead or contact.

fold line The bend point of a flexible circuit substrate.

footprint pads See land pattern.

footprint pattern See land pattern.

functional testing Simulation testing of a finished circuit assembly.

glass reinforced resin laminate Typical dielectric base for rigid circuit substrates of epoxy resin and glass fiber.

grid pattern A uniform space between features on a substrate usually referring to via or feedthrough hole patterns.

grid position Location of device or hole center location from a zero reference or datum point.

gull wing Contact lead formed downward and away from the device body in a shape similar to the wing of a gull.

heat dissipation Thermal transfer of heat from the circuit assembly.

heat pressure seal A method for terminating conductive ribbon cable to a mating conductive pattern on a substrate.

HSC Heat Seal Connection.

in-circuit-test The active test of each device on the assembled circuit substrate through probe contact by a test fixture.

induction heating Heating transferred by blower or fan.

infrared Frequency of light spectrum used to locally generate energy or heat to reflow or liquefy solder paste alloy.

interconnection The electrical interface of conductor circuits with electronic devices.

interface junction Electrical and mechanical joining point.

IPC The Institute for Interconnection and Packaging of Electronic Circuits.

IPC-SM-782 Land pattern standards for surface-mounted devices.

IR See infrared.

J-lead contacts Device leads formed in the shape of a ''J'' at the contact point of the substrate land pattern.

JEDEC Joint Electronic Device Engineering Council.

land pattern The geometric contact shape provided on the substrate surface for the electrical and mechanical interface of surface-mounted devices.

LCD Liquid Crystal Display.

lead spacing See center-to-center spacing.

leaded devices Electronic components with leaded contacts for pin-through-hole termination with the substrate.

liquid photo-imaged polymer Material for solder mask on rigid substrate circuit boards.

liquid solder A tin-lead alloy heated to a liquid point for the electrical termination of devices to a substrate.

magazine A rectangular tube carrier for storing and handling of IC devices.

MELF Metal Electrical Face device, cylindrical shape with end-cap contact terminals.

migration of liquid solder   The flow of the liquid solder alloy during processing.

MMD/MMT   Micro Miniature Diode and Micro Miniature Transistor.

modules   A circuit subassembly.

multilayer   Construction of three or more conductor trace layers separated by dielectric materials to form a circuit substrate.

necking down   Reducing the conductor trace width from a wide to narrow cross section.

net   Common electrical connection between two or more device contacts or leads.

node   Common point of contact for testing a circuit assembly.

networks   Two or more devices interconnected in a single component package.

nickel barrier   Plated nickel alloy on the end cap contacts of passive devices to prevent the migration of silver palladium termination material.

offset stepping   Staggering of circuit traces or via hole pads.

oval extension   Material added to hole land patterns for added strength or to prevent drilled hole break-out on small via interconnections.

pad geometry   See land pattern.

panel   A substrate with specific features furnished for direct assembly machine processing.

panel format   See panel.

panelization   Combining two or more substrates on a single panel for direct assembly machine processing.

partition   Separation of circuit functions or power and ground area.

partitioned cluster   Electrical functions separated to accommodate isolated testing of circuit parts.

pick-and-place assembly equipment   See automated placement.

pin and socket connectors   Pin interface to a mating socket for the electrical connection of assemblies or modules.

pin and socket headers   See pin and socket connectors.

pin and socket strips   See pin and socket connectors.

PLCC   Plastic Leaded Chip Carrier IC package with J-lead contacts.

polarized parts   Devices that must be mounted in a predetermined direction or orientation.

polyimide glass   Polyimide resin and glass fiber substrate material.

polyimide quartz   Polyimide resin and quartz fiber substrate material.

polymer coating   Material used for solder masking on substrates.

polyimide   A stable resin film used for the base material of flexible substrate circuits.

potentiometers   Variable or adjustable resistor.

preblanked units   Die cut or machined substrates.

preformed leads   Leads prepared for direct assembly processing.

process friendly   Compatible materials in the assembly processing of surface-mounted devices to a circuit substrate.

profiling   Temperature adjustment for the reflow processing of solder paste for electrical and mechanical interconnection of SMT devices to the substrate.

prototype   Limited quantity assembly processing of a product.

PTH   Pin-through-hole leaded device or component.

punch and retain   Individual substrate units die punched from and returned to a panel for assembly processing.

punch cut   See die cut.

punch press   Machine for punch press processing or profiling of materials.

punching   See die cut.

QFP   Quad flat pack IC device with component leads extending from each of the four sides.

rat's nest technique   Preliminary interconnect planning for electrical circuits.

rectifiers   See diode.

reflow   The process of heating solder paste to a liquid form to accomplish electrical and mechanical interface of components to the substrate.

reflow assembly   See reflow.

reflow solder   See reflow.

removal system   Tools or equipment designed for the removal of surface-mounted devices.

resistor   A device for limiting or restricting electrical current flow in a circuit.

retention mounting tabs or bosses   Mechanical features for mounting or strengthening large components or connectors.

rework   Correcting defects that do not meet the established quality requirement specified for the product.

right-angle mounting   Component or device designed to mount at a ninety degree rotation from the substrate surface.

rigid circuits   Nonflexible substrate with one or more conductive circuit layers for interconnecting electronic devices.

rigid substrate   See rigid circuits.

robotic assembly   See automated placement.

robotic equipment   Automated machines for the assembly of electronic products.

routing template   Profile guide for the finished shape of a rigid substrate.

SRAM   Static read and write accessible memory device.

secondary assembly   Operations following the primary assembly process.

secondary hot air process   Reflow of solder with heated air or gas for mounting devices following the primary assembly process.

secondary wave solder   The soldering of leaded and epoxy attached devices in a liquid wave of molten solder alloy.

self-centering   An alignment to the center location on the land pattern during the reflow-solder process.

side loading   Stress introduced to the component or substrate from one or more sides.

SIMM   Single in-line memory module.

SIP   Single in-line pin (leaded) module.

small outline   Miniature surface-mounted integrated circuit package.

SMD   Surface-mounted device.

SMEMA   Surface Mount Equipment Manufacturer's Association.

SMOBC   Solder mask over bare copper.

SMT   Surface Mount Technology.

SMTΛ   Surface Mount Technology Association.

SOIC   See small outline.

SOJ-IC   Small outline IC with J-lead configuration.

solder connection   Termination of device contacts to the substrate circuit with tin-lead alloy.

solder bridging   See bridging.

solder buildup   Excessive solder at the connection of a device to the substrate circuit.

solder fillet   The tin-lead junction interconnection of the surface-mounted component contact to the land pattern of the substrate.

solder mask   Dielectric material applied to the substrate surface for masking of circuit traces during solder processing.

solder mask barrier   Applies to the solder mask material reserved for containment of solder paste on land patterns during the reflow process.

solder migration   See migration.

solder paste   A tin/lead alloy powder mixed with flux, solvent, and binders for application to SMT land patterns on the circuit substrate.

solder plating   Electroplating of solder alloy to copper conductors or features on a substrate.

solder screen   A pattern masked steel mesh used to deposit solder paste to contact land patterns on the substrate surface.

SOP   Small outline IC package (Japan).

SOT   Small outline transistor/diode package.

SOT-23   Three-lead, small outline transistor/diode package.

SOT-24   Four-lead, small outline transistor/diode package.

SOT-89   Higher power surface-mounted transistor package.

space analysis   Calculation of component area requirement and circuit density.

spring action   Automatic retraction or return to normal position.

spring loaded test probes Test probe contacts designed to mate with leads or contact points provided on the circuit substrate surface.

stepped solder Solder paste deposited in multiple thickness to accommodate multistep soldering of selected devices.

strain relief mounting tabs or bosses See retention mounting tabs or bosses.

structural substrate See molded substrate.

surface-mounted packages See surface-mounted devices.

surface-mounted technology The materials and process for contact attachment of electronic devices directly to the substrate surface.

TAB devices Tape automated (lead frame to silicon die) bonding process for low cost and high volume IC packaging.

tab retaining points A small cross section of the base material for the retention of individual substrate units in a panel format.

tantalum capacitor A polarized electron (voltage) storing device having a tantalum dielectric for higher voltage applications.

tape and reel A continuous strip packaging furnished on a reel carrier for direct assembly machine processing.

TCE The linear thermal expansion per unit change in temperature (often referred to as CTE).

tear restraints Copper pattern retained in the corners of a flexible circuit substrate.

tent over vias The covering of via hole and pads by solder mask material.

termination area See land pattern.

test pad Contact area designated for probing by automatic test systems.

test probes See spring loaded test probes.

thermal coefficient See TCE.

thermal cycles Time between high and low temperature levels during static or dynamic test.

TO-5 Multileaded metallic case transistor package.

TO-92 Multileaded plastic case transistor package.

tombstoning The lifting of one side or end of a device from the substrate surface during reflow-solder processing.

tooling fixtures See fixtures.

tooling holes Nonplated holes in the substrate used for alignment to machines and fixtures.

tooling pins Alignment posts that mate with tooling hole in the substrate.

touch-up tools Solder tools designed for safe rework of solder joint connections.

trace width The dimensional width of a conductor on substrate layer.

transistor Discrete semiconductor for current switching and signal amplification.

tray carriers Partitioned packaging furnished to transport quad flat pack IC devices.

tube See magazine.

tube magazine See magazine.

universal fixture Tooling designed for use on several differently shaped substrate assemblies.

UV cured epoxy See epoxy.

vapor phase A method of reflowing solder paste in a vapor created by boiling a unique chemical solution.

via pad A plated hole used to interconnect two or more circuit layers on a substrate.

wave solder Soldering of leaded and epoxy attached surface-mounted devices by a controlled contact with a machine generated wave of molten solder alloy.

Index

S

T